T0076156

Electrical Engineering Developments

ADVANCED CONTROL DESIGN OF MEMS VIBRATORY GYROSCOPES

Nanotechnology, Science and Technology

Additional books in this series can be found on Nova's website under the Series tab.

Additional E-books in this series can be found on Nova's website under the E-book tab.

Electrical Engineering Developments

Additional books in this series can be found on Nova's website under the Series tab.

Additional E-books in this series can be found on Nova's website under the E-book tab.

Electrical Engineering Developments

ADVANCED CONTROL DESIGN OF MEMS VIBRATORY GYROSCOPES

JUNTAO FEI
EDITOR

Nova Science Publishers, Inc.
New York

For permission to use material from this book please contact us:
Telephone 631-231-7269; Fax 631-231-8175
Web Site: http://www.novapublishers.com

Additional color graphics may be available in the e-book version of this book.

Library of Congress Cataloging-in-Publication Data

Advanced control design of MEMS vibratory gyroscope / editor, Juntao Fei.
p. cm.
Includes bibliographical references and index.
ISBN 978-1-61470-487-4 (hardcover : alk. paper) 1. Gyroscopes. 2. Adaptive control systems. 3. Accelerometers. 4. Microelectromechanical systems. I. Fei, Juntao.
TJ209.A33 2011
681'.753--dc23

2011024359

Published by Nova Science Publishers, Inc. † New York

CONTENTS

PREFACE

MEMS (Micro Electro Mechanical Systems) technologies were developed by applying semiconductor microfabrication technologies, making three-dimensional microstructures and mechanical systems. MEMS technologies offer the advantages of a batch fabrication of numbers for devices and the ability to integrate multiple functional units in a small area; the latter being important for developing smart and sophisticated devices. Gyroscopes are commonly used sensors, used for measuring angular velocity in a wide variety of applications, including: navigation, homing, and control stabilization. Fabrication imperfections and thermal, mechanical noise may hinder the measurement of angular velocity with the MEMS gyroscope. The angular velocity measurement and minimization of the cross coupling between two axes present challenging problems when controling MEMS gyroscopes.

This book begins with an overview of concepts from the adaptive control system theory. Furthermore, this book presents a comprehensive treatment of the analysis and advanced control design of the MEMS gyroscope, addressing the problem of angular velocity measurement and a minimization of the cross coupling between two axes. First, the dynamical modeling procedure is described. The book then continues to describe how control design moves from an adaptive control of the MEMS gyroscope to a robust adaptive control of the MEMS gyroscope; a progression which culminates in a robust adaptive sliding mode control hierarchy, not requiring upper bound information of external disturbance. Additionally, novel advanced control methods are presented: adaptive control and robust adaptive control optimization in state tracking and parameter estimation, adaptive sliding mode control for tracking of reference systems and parameter convergence analysis, robust adaptive sliding mode control, and adaptive sliding mode observer design, adaptive neural network sliding mode control. These methods generate solutions of guaranteed performance for problems of robustness, Lyapunov stability, and parameter convergence. Examples, simulations, and comparative studies to the fundamental issues all serve to illustrate the technical approaches and verify the performances of the various advanced control designs.

This book is organized into 12 chapters:

Chapter 1. Introduction to MEMS Gyroscope
Chapter 2. Dynamics of MEMS Gyroscope
Chapter 3. System Identification of MEMS Gyroscope
Chapter 4. Adaptive Control of MEMS Gyroscope
Chapter 5. Robust Adaptive Control of MEMS Gyroscope
Chapter 6. Adaptive Sliding Mode Control of MEMS Gyroscope
Chapter 7. Robust Adaptive Sliding Mode Control of MEMS Gyroscope
Chapter 8.Adaptive Sliding Mode Control and Observer of MEMS Gyroscope
Chapter 9.Comparative Study of Adaptive Control of MEMS Gyroscope
Chapter 10. Robust Adaptive Sliding Mode Control of MEMS Triaxial Gyroscope
Chapter 11. Neural Network Based Adaptive Sliding Mode Control of MEMS Gyroscope
Chapter 12. Conclusions

Chapter 1 gives an introduction to the MEMS gyroscope and briefly points out the major contributions of this book. It goes on to compare different areas of control theory and introduce some basic concepts of adaptive control. Chapter 2 describes the fundamentals and dynamics of MEMS gyroscopes. The non-dimension version of the dynamics of the MEMS gyroscope is also derived.

Chapter 3 presents a novel adaptive approach that can identify, in an on-line fashion, angular rate, as well as other system parameters. By rewriting the dynamic model of MEMS gyroscope sensor that can update the estimator of angular rate adaptively and converge to its true value asymptotically, the proposed approach develops a series-parallel on-line identifier. The feasibility of the proposed approach is analyzed and proved by Lyapunov's direct method.

Chapter 4 proposes an adaptive control approach for a MEMS z-axis gyroscope sensor. The dynamical model of the MEMS gyroscope sensor is derived and an adaptive state tracking control for the MEMS gyroscope is developed. The proposed adaptive control approaches can estimate the both angular velocity and the damping and stiffness coefficients, including the coupling terms due to the fabrication imperfection. The stability of the closed-loop systems is established with the proposed adaptive control strategy.

Chapter 5 develops a robust adaptive controller for the MEMS gyroscope. The robust adaptive control algorithm can estimate the angular velocity vector and the linear damping and stiffness coefficients in real time. A robust adaptive controller, incorporating model uncertainties and external disturbances is derived in the Lyapunov sense and the stability of the closed-loop system is established. The numerical simulation is presented to verify the effectiveness of the proposed control scheme.

Chapter 6 presents an adaptive sliding mode tracking controller with a proportional and integral switching surface. A novel adaptive sliding mode controller, based on model reference adaptive state feedback control, is proposed to deal with the tracking problem for a class of dynamic systems. Instead of a conventional sliding surface, a proportional and integral sliding surface is chosen and then a class of an adaptive sliding mode controller with integral sliding term is developed. It is shown that the stability of the closed-loop system can be guaranteed with the proposed adaptive sliding mode control strategy. This chapter also presents a new adaptive sliding mode controller with a proportional and integral sliding surface for the MEMS gyroscope. An adaptive sliding mode controller that incorporates both matched and unmatched uncertainties and disturbances is derived and the stability of the closed-loop system is established. The adaptive sliding mode control algorithm can estimate the angular velocity vector and the linear damping and stiffness coefficients in real time.

Chapter 7 derives an adaptive sliding mode controller for a MEMS z-axis vibratory gyroscope with unknown upper bound estimation in the presence of model uncertainties and external disturbances. The adaptive model references state tracking control with a sliding mode controller is developed to estimate the angular velocity and the damping and stiffness coefficients of a MEMS vibratory gyroscope in real time. An adaptive controller that can estimate the unknown upper boundaries of the parameters with uncertainties and external disturbances is derived and the stability of the closed-loop system is established. The adaptive sliding mode controller is designed so that the trajectory of the driving axes can track the state of the reference vibration model with unknown upper bound of model uncertainties and external disturbances.

Chapter 8 investigates a novel adaptive sliding mode control with a sliding mode observer for a MEMS gyroscope. In the presence of parameter variations and external disturbances, the proposed adaptive sliding mode controller, with a sliding mode observer, can estimate the unmeasured velocities, angular velocity and damping and stiffness coefficients of the gyroscope. A novel adaptive sliding mode controller with a proportional and integral sliding surface is derived and the stability condition of the closed-loop feedback system is established.

Chapter 9 conducts a comparative study of adaptive control approaches for MEMS z-axis gyroscopes. Novel adaptive controllers and adaptive sliding mode controllers are proposed, respectively, and the comparative analysis of these two methodologies is implemented. The difference in the controller derivation and stability analysis is also discussed in detail. The proposed adaptive control approaches can estimate the angular velocity and the damping and stiffness coefficients, including the coupling of terms. The stability of the closed-loop systems are established with the proposed adaptive control strategies.

Chapter 10 presents a robust tracking control strategy, using an adaptive sliding mode approach for a MEMS triaxial angular sensor device, which is able to detect rotation in three orthogonal axes by using a single vibrating mass. An adaptive sliding mode controller with proportional and integral sliding surface is developed and the stability of the closed-loop system can be guaranteed with the proposed adaptive sliding mode control strategy. The proposed adaptive sliding mode controller updates estimates of all stiffness errors, damping and input rotation parameters in real time; removing the need for any offline calibration stages. To enable all unknown parameter estimates to converge to their true values, the necessary model trajectory is shown to be a three-dimensional Lissajous pattern.

Chapter 11 proposes a robust adaptive sliding mode control strategy of the MEMS triaixal gyroscope, using a radial basis function (RBF) network. The adaptive RBF neural network is incorporated into the adaptive sliding mode control in the same Lyapunov framework and the stability of the proposed adaptive neural sliding mode control can be established. A key property of this scheme is that prior knowledge of the upper bound of the system uncertainties is not required. An adaptive RBF neural network is used to learn the unknown upper bound of model uncertainties and external disturbances. Additionally, the proposed adaptive sliding mode controller updates estimates of all stiffness errors, damping terms and angular velocities in real time.

Chapter 12 presents the concluding remarks and suggests a list of theoretical and practical topics for further research in this area of advanced control of the MEMS gyroscope.

The book will have a large potential readership; ranging from scientists and students, crossing over many disciplines; to mechanical and electrical engineers. This book is an essential reading for both engineers and scientists working in the fields of micro electro-mechanical systems, advanced control, and related MEMS technologies. It will also be of interest to senior undergraduate and graduate students; as well as aerospace, mechanical, electrical design engineers who want to acquire some background in advanced control design of the MEMS gyroscope. On the scientific plane, the book will provide important new information to mechanical engineers, electrical engineers and scientists. The new data and insights presented here will enable students to have a better grasp of the complicated advanced controls, as applied to the MEMS vibratory gyroscope. In short, this book will undoubtedly serve as a milestone of the integrative approaches that are needed to better comprehend and manage the emerging MEMS world.

I wish to gratefully acknowledge the valuable help rendered by institutions and individuals throughout the conduction of the research presented in this book. This research was partially supported by the National Science Foundation of China under Grant No. 71074057, The Natural Science Foundation of Jiangsu Province under Grant No.BK2010201, Scientific Research Foundation for the Returned Overseas Chinese Scholars, State Education Ministry, Scientific Research Foundation of High-Level Innovation and Entrepreneurship Plan of Jiangsu Province, NASA Grant No. NCC5-573, and the Governor's IT Initiative in the University of Louisiana. We would like to thank their financial support that made this research possible. First, my sincerest thanks to Professor Celal Batur at The University of Akron. He offered me exceptional working conditions over the years that I was under his supervision. For giving me the opportunity to do research on the fascinating topics covered in this book, I truly appreciate his constant support in so many different aspects, including his supervision on Chapter 2 and Chapters 5-8. I am also thankful to Hohai University, the University of Akron, and the University of Louisiana at Lafayette for a pleasant and supportive research environment. I also extend my thanks to Professor Fahmida Chowdhury, for her support when I worked in the University of Louisiana; Mr. Yuzheng Yang, for his contribution to Chapter 3; and Mr. Hongfei Ding, for his contribution to Chapter 11. Finally, I am especially grateful to my family and for their support to my research work, which made this project possible.

Chapter 1

INTRODUCTION TO MEMS GYROSCOPES

1.1. INTRODUCTION

Gyroscopes are commonly used sensors for measuring angular velocity in many applicational areas, such as in the stability and navigation control in spacecrafts and airplanes, and rollover detection for automotives, consumer electronics, robotics, etc. Gyroscopes, through the Coriolis acceleration, are the devices that transfer energy from one axis to another axis. The operating principle of the vast majority of all existing micromachined vibratory gyroscopes relies on the generation of a sinusoidal Coriolis force, due to the combination of a vibration of a proof-mass and an orthogonal angular-rate input. The proof mass is generally suspended above the substrate by a suspension system consisting of flexible beams. The overall dynamical system is a typical two degrees-of freedom (2-DOF) mass-spring-damper system; wherein the rotation-induced Coriolis force causes the energy transfer to the sense-mode, proportional to the angular velocity input.

Micromachined gyroscopes can be a potential alternative to expensive and bulky conventional inertial sensors. High-performance gyroscopic sensors, including precision fiber-optic gyroscopes, ring laser gyroscopes, and conventional rotating wheel gyroscopes, are too expensive and too large for use in most emerging applications. With the micromachining process, a micro-sized gyroscope will provide high accuracy rotation measurements. Thus, miniaturization of vibratory gyroscopes with innovative micro-fabrication processes and gyroscope designs is expected to become an attractive solution to current inertial sensing market needs, as well as open new market opportunities.

Recent advances in micro-machining technology have made the design and fabrication of MEMS gyroscopes possible. These devices are several orders of magnitude smaller than conventional mechanical gyroscopes, and they can be fabricated in large quantities by batch processes. With their dramatically reduced cost, size, and weight; MEMS gyroscopes potentially have a wide application spectrum in the aerospace and automotive industries, military, and consumer electronics markets. The automotive industry applications are diverse, including: high performance navigation and guidance systems; ride stabilization; advanced automotive safety systems, such as yaw and tilt control; roll-over detection and prevention; GPS augmentation, such as MEMS inertial navigation sensor embedded GPSs; and next generation airbag and anti-lock brake systems. The military applications relate to micro airplanes, satellite controls, etc. A wide range of consumer electronics applications include: image stabilization in video cameras, video games, virtual reality products, and a 3D mouse; inertial pointing devices; and multiple purposes throughout the computer gaming industry.

1.2. Control of MEMS Gyroscopes

The emergence of MEMS technology is opening up new market opportunities and applications in a wide variety of areas. One of the most important emerging fields in MEMS design is the control-methodology-based MEMS. This approach utilizes feedback control to compensate for fabrication/design flaws and, thus, ensures the system's robust performance. In this approach, not only a control algorithm is integrated into a system, but the MEMS structure design is also altered accordingly.

Control refers to forcing a system to behave in a desired manner. The integration of control systems and MEMS is becoming more common since VLSI logic circuits can be fabricated on the same substrate as the mechanical elements. The role that control plays in MEMS is varying, some including:

- Providing precise control over MEMS actuation elements as electrostatic plates.
- Improving response time and device input range and accuracy through force-feedback schemes.
- Providing real-time estimates of the unknown parameters of a MEMS system.

- Compensating for the effects of fabrication imperfections.
- Compensating for time varying effects.

There are some noises, such as sensing noise or noise existing in the various angular velocity sensors. Noise can enter at various stages of the position and/or velocity sensing process, from the sensing element itself or from any associated sensing circuits and wiring. The most common noise source is parasitic capacitances that form between capacitive sensing elements. Significant improvements have been found by using an insulating substrate, such as glass. Phase differential sensing schemes have been shown to be robust against variations in sensing element scale factors [50], [57] and [72]. An alternative tunneling-based sensing element, only for angular velocity sensors operating in the force-balancing mode, has been employed to improve mass displacement sensitivity [71].

Due to the temperature dependence of Young's Modulus and thermally induced localized stresses, variations in the temperature of the structure also perturb the dynamical system parameters. Thermal mechanical noise appears as a white noise force, acting on the vibrating proof mass wherever there is damping. Thermal mechanical noise limits the ultimate achievable resolution of all vibratory angular velocity sensors. A practical correlation filter can greatly reduce thermal noise in a gyroscope's output signals, compared to using a conventional low pass filter [77]. Another approach to de-noising gyroscope signals is presented by Wang and Huang, using wavelet packet analysis [70].

An angular velocity sensor that operates in an open-loop mode employs no feedback control in either the drive or sense axes. The sensitivity is increased if the drive and sense axes have matched resonant frequencies, due to mechanical amplification in the Coriolis axis. Any mismatch in the resonant frequencies between the drive and sense axes can be tuned by applying a force to offset to one of the axes. This has the effect of altering its stiffness. Alper and Akin [53] presented an open loop angular velocity sensor with a symmetric structure that provided matched and decoupled resonant modes.

Most angular velocity sensors by Mochida et al. [74] and Alper and Akin [53] employ symmetric structures to reduce any asymmetries in the device dynamics. Asymmetric stiffness causes unwanted forces to act on the mass, forces the proof mass to trace a circular path instead of a linear one, resulting in a zero rate output (ZRO) called quadrature error [73]. ZRO is an erroneous output signal from the device when it is subject to zero angular velocity input. One way to reduce the ZRO is to employ mismatched modes to uncouple the two axes and better isolate the driving motion from the piezoresistive sensing elements. The other solution to increasing bandwidth is adding a negative feedback loop which nulls the motion of the mass in the sense axis. Acar et al. [51] use two coupled masses; each with two degrees of freedom and separate resonant peaks, operating the device in the flatter and wider operating region between the two peaks for more robust performance. This sort of approach was further developed by Acar and Shkel [52] in a ring arrangement, in which eight interconnected vibrating masses are used; resulting in a very wide bandwidth that is robust against temperature fluctuations and has a small ZRO, due to the decoupling nature of the multiple mismatched modes. Differential phase demodulation has been shown to be insensitive to variations in drive amplitude [50], [57] and [72]; however, its use is limited to open-loop operation. Oboe et al. [77] describes the design of the control loops in a z-axis, MEMS vibration gyroscope operating in a vacuum enclosure.

Past MEMS gyroscope research focused on the development of the micro-electromechanical fabrication processes, sensor modeling, and the active control of the sensor dynamics for model identification and angular velocity sensing [14], [15], and [17]. Several control algorithms have been proposed to control the MEMS gyroscope. In terms of automatic control, a force-balancing feedback control scheme, using sigma-delta modulation [37], has been proposed as a conventional mode of operation. In the force-balancing mode [77], instead of using the sense axis motion resulting from the Coriolis force as output, it is used as an input to a negative feedback loop. Using the control signal, this feedback loop acts to nullify the motion in the sense axis. The demodulated feedback force then becomes the output of the device: it is proportional to the angular velocity input in an ideal system. More complex control strategies are required to separately identify and compensate for the errors and the angular velocity. The force-balancing mode is an integral part of the tunneling- based angular velocity sensor that was proposed by Kenuba et al. [71],using a force-balancing mode to maintain the tunneling gap.

Leland et al. [19] proposed one adaptive controller which estimates both the cross stiffness and input angular velocity; compensating for their effects on the sense axis in a force balancing mode. Dong et al. [84] proposed a novel active disturbance rejection control by an extended state observer to estimate the rotation rate of the MEMS gyroscope. Zheng et al. [85] introduced a novel oscillation controller for the drive axis of a MEMS gyroscope that is a traditional PD controller, as well as a linear extended state observer. Since the controller design does not require exact information for system parameters, it is very robust against structural uncertainties of the drive axis.

Park and Horowitz [14] presented adaptive add-on control algorithms for the conventional mode of operation of MEMS z-axis gyroscopes. This scheme is realized by adding an outer loop to a conventional force-balancing scheme that includes a parameter estimation algorithm. The parameter adaptation algorithm estimates the angular velocity and identifies and compensates the quadrature error.

Adaptively controlled systems can be characterized as those that can adapt to changes in the system dynamics over time. The adaptive mode of operation is better suited for medium-cost gyroscopes that are used in high-performance applications.

Leland [17] proposed an adaptive controller for an open loop device that tunes the drive axis's frequency of vibrations to match that of a fixed reference drive signal. Since the dynamics of the system are dictated by the time invariant reference model instead of the natural resonant frequency of the device (which may be time varying); this offers advantages over using a Phase locked loop (PLL), which tunes the driving signal frequency to match the resonant frequency. Leland [18] added adaptive control to regulate the drive oscillation amplitude; compensate for quadrature error, due to asymmetric stiffness; and nullify the sense axis vibration in the force-balancing mode. Leland et al. [17] present an adaptive controller for the drive axis of a vibration gyroscope. By using a destabilizing positive feedback with automatic gain control, a piezoelectrically driven gyroscope is operated as an oscillator circuit. Recently, Leland et al. [19] incorporated a time varying input rate into a Lyapunov based adaptive controller; estimating and compensating for both input angular velocity and cross stiffness terms, by using a polynomial approximation techniques.

Asymmetric stiffness and damping arise from fabrication imperfections. It is difficult to separate the Coriolis signal from the asymmetric damping signal, and Park [22] has been able to distinguish one from another by using a model reference adaptive control approach. Closkey et al. [73] used off-line lattice-filter based algorithms to estimate high-order linear MIMO models for their cloverleaf vibrating post angular velocity sensor dynamics. All the stiffness terms were successfully identified; however, it was impossible to identify the cross damping terms, as they are indistinguishable from the angular velocity terms. Later work on the same cloverleaf design proposed evolutionary optimization computation to tune the mismatch in resonant modes of an off-line fashion [59].

Salah et al. [74] proposed a new adaptive controller to control both axes of a z-axis MEMS gyroscope and to facilitate a time varying angular velocity sensing. First developed is an off-line adaptive least square estimation strategy, accurately estimating the unknown model parameters. An online active controller/observer is then developed for time-varying angular velocity sensing.

By providing the Coriolis force as the input to the adaptive estimator, Feng et al. [87] presented an adaptive estimator-based technique to estimate the angular motion and to improve the bandwidth of microgyroscopes. Jagannathan et al. [21] presented an adaptive force-balancing control (AFBC) scheme with actuator limits for a MEMS Z-axis gyroscope. The proposed AFBC scheme controls the vibratory modes of the proof mass, while still ensuring that the control input satisfies the magnitude constraints and enhances the performance of the gyroscope in the presence of fabrication uncertainties. Implemented using digital processors, Park and Horowitz [24] presented a discrete time version of the observer-based adaptive control system for micro-electro-mechanical systems gyroscopes..

Sliding mode control is a robust control technique which has many attractive features, such as robustness to parameter variations and insensitivity to disturbances. The sliding mode controller is composed of an equivalent control part that describes the behavior of the system when the trajectories stay on the sliding manifold. Simultaneously, a variable structure control enforces the trajectories to reach the sliding manifold and prevent them leaving the sliding manifold. It has some limitations in practical applications, such as chattering or high frequency oscillation. Adaptive control is an effective approach to handle parameter variations. Adaptive methods are used to automatically adjust the response of the controller, compensating for changes in the response of the plant. Therefore, adaptive sliding mode control has the advantages of combining the robustness of variable structure methods along with the tracking capability of adaptive control. In the adaptive sliding mode control, sliding mode control is combined with a linear controller, adjusted by adaptive law. The tracking error between the plant state and reference model is fed back to the sliding mode and adaptive controllers.

Utkin [1] and [2] introduced the variable structure system and showed that variable structure control is insensitive to both parameter perturbations and external disturbances. Narendra et al. [3], Astrom et al.[4], Ioannou and Sun [5], and Tao [6] described the model reference adaptive control. In the last few years, many applications have been developed using sliding mode control and adaptive control. The adaptive control law, merging parameter identification and sliding mode control, was proposed and analytically studied by Fradkov and Andrievsky [79-82]. Lee et al. [7] developed a variable structure augmented adaptive controller for a gyro platform. Wang and Sinha [8] presented an adaptive sliding mode controller for a microgravity isolation system. Song and Mukherjee [9] proposed a smooth robust compensator. Sam et al. [10] presented a class of proportional and integral sliding mode controls, specifically applying to an active suspension system. Chouand Cheng [11] and Lin et al. [12] proposed an integral sliding surface and derived an adaptive law to estimate the upper boundary of uncertainties. Hsu et al. [13] developed a design of input/output based variable structure adaptive control. Adaptive laws to estimate the upper boundary of uncertainties and disturbances are discussed in [37-49].

Batur et al. [20] developed a sliding mode control for a MEMS gyroscope system. A model reference adaptive feedback controller and a sliding mode controller have been considered to guarantee the stability of the MEMS device. Park [22] proposed an adaptive controller for two-axes driven MEMS gyroscope, driving both axes of vibration and also controls the entire operation of the gyroscope. Park and Horowitz [23] proposed an adaptive control scheme which estimates the component of the angular velocity vector and compensates for friction forces and fabrication imperfections. John and Vinay [58] extended Park's method [22] to triaxial angular sensors and presented a novel concept for an adaptively controlled triaxial angular sensor. A surface-micromachined dual axis gyroscope, based on rotational resonance of a 2 μm thick polysilicon rotor disk, has been reported in [22]. Since the disk is symmetric in two orthogonal axes, the sensor can sense rotation equally about two axes. This is suitable for an adaptive mode of operation, driving both axes while allowing equal movements in the drive and sense axes..

In practical MEMS gyroscopes, the axial velocities may be immeasurable. The use of velocity sensors may add to the cost, size, and weight of the vibrating platform. It is necessary to design an observer to estimate the unmeasured states. A high-gain observer was investigated by Esfandiari and Khalil [25] for the design of an output feedback controller, due to its ability to robustly estimate the unmeasured states while also asymptotically attenuating disturbances. Krstic et al. [27] proposed an adaptive backstepping nonlinear observer for a nonlinear system. The basic sliding mode observer structure consists of switching terms that were added to a conventional Luenberger observer. The sliding mode design method enhances robustness over a range of system uncertainties and disturbances. Slotine et al. [27] and Walcott and Zak [28] designed sliding mode observers with an additional switching function, addressing plant uncertainties by using the Lyapunov stability theorem. A sliding mode controller and observer for a microgravity isolation system was investigated in [29]. Batur and Zhang [30] proposed a sliding mode observer and controller design for a hydraulic motion control system. Edwards and Spurgeon [31-33] described the sliding mode observer theory and application examples. A class of nonlinear extended state observers was proposed by J. Han [34] as a unique observer design. The latter is rather independent of the mathematical model of the plant, thus achieving inherent robustness. A number of robust state observers have the form of the standard observer, plus a strengthening function to reflect the unknown perturbation, such as model uncertainty and external disturbance. Moura et al. [35] suggested a function to estimate plant uncertainties and added it to the state estimator.

1.3. Original Contribution of the Book

The design and fabrication of MEMS gyroscopes has been the subject of extensive research over the past few years. Moreover, advances in the fabrication techniques allow both detection and control electronics to be integrated on the same silicon chip, together with the mechanical sensor elements. The cost of MEMS gyroscopes is decreasing, while their accuracy is continuously being improved. There are certain factors which may affect the performance of gyroscopes; these are known as the coupling parameters. Coupling parameters are the terms which exist in the drive and sense motion equations in the form of stiffness and damping terms; multiplied by position and velocity states, thereby relating the drive and sense motion equations. Coupled damping parameters are usually small and are not so important, compared to the coupled stiffness terms.

Most angular velocity sensors employ symmetric structures to reduce any asymmetries in the device dynamics. However, due to the limitations of fabrication, it is unlikely that the principle stiffness axes or damping axes will be perfectly aligned with the geometric axes of the device. The asymmetric stiffness terms may arise from several circumstances, including a situation whereby the centre of the mass may not perfectly coincide with the geometric centre or when the supporting springs may have unequal stiffness. The quadrature error forces are proportional to the mass displacement and are, therefore, 90 degrees out of the phase (quadrature phase) with the Coriolis forces, which are proportional to the mass velocity. The quadrature phase makes asymmetric stiffness forces distinguishable from Coriolis forces. Like asymmetric stiffness, asymmetric (or cross) damping arises from fabrication imperfections. It is the misalignment of the principle damping axes from the geometric axes of the device. It also causes erroneous forces to act on the mass; however, they are in the phase with the Coriolis force since they are both proportional to the mass velocity. By demodulating, with respect to both a sine and a cosine signal, the quadrature error can be distinguished from the combined Coriolis and asymmetric damping signals.

Chapter 2

Dynamics of MEMS Gyroscopes

2.1. MEMS Vibratory Angular Velocity Sensors

Prior to the emergence of the MEMS angular velocity sensors, the conventional devices for measuring angular velocity include: the spinning mass gyroscope, ring laser gyroscope and fibre optic gyroscope. Recently, great efforts have gone towards the research of MEMS angular velocity sensing, resulting in their substantial magnitude of performance improvements every two years.

While there are many different structural configurations that have been used for MEMS angular velocity sensing, including: vibratory masses, rings, stars, tuning forks, posts, beams, and butterflies; they all operate using the same basic principle. The structure is driven into oscillation in a primary mode of vibration; when the device is subjected to an angular velocity, energy from the primary mode is transferred to a secondary mode which causes it to oscillate. This transfer of energy to the secondary mode is due to the Coriolis effect and is indicative of the angular velocity input.

The MEMS vibratory gyroscope is shown in Figure 2.1, including drive axis displacement, the angular velocity input, and the resulting sensing axis motion. The physical structure consists of two degrees of freedom (DOF) spring mass damper system. The proof mass is driven into oscillation in one axis and the mass' displacement is sensed in a perpendicular axis. This perpendicular vibration is caused by a transfer of energy from the primary to the secondary vibration axis, through a rotation induced Coriolis force, acting on the mass. Typically for MEMS devices, the motion in the sense axis is an order of magnitude smaller than that of the drive axis. Next, we will discuss the physical structure and mechanical and electrical design of MEMS sensing and actuation.

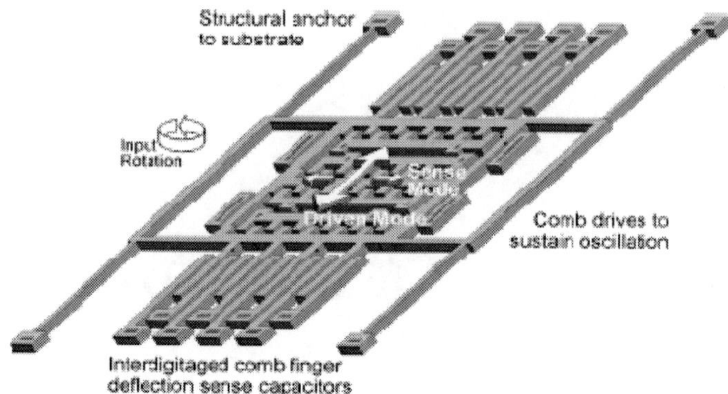

Figure 2.1. Vibratory MEMS gyroscope.

2.2. Mechanical Design

All existing micromachined vibratory gyroscopes operate on the principle of detecting the sinusoidal Coriolis force, induced on a vibrating proof-mass in the presence of an angular-velocity input. Since the induced Coriolis force is orthogonal to the drive-mode vibration, the proof-mass is required to be free to oscillate in two orthogonal directions, and is desired to be constrained in other vibration modes. The proof-mass is generally suspended above the substrate by a suspension system consisting of thin flexible beams, usually formed in the same structural layer as the proof-mass. The guided-end cantilever beam is shown in Figure 2.2 [83].

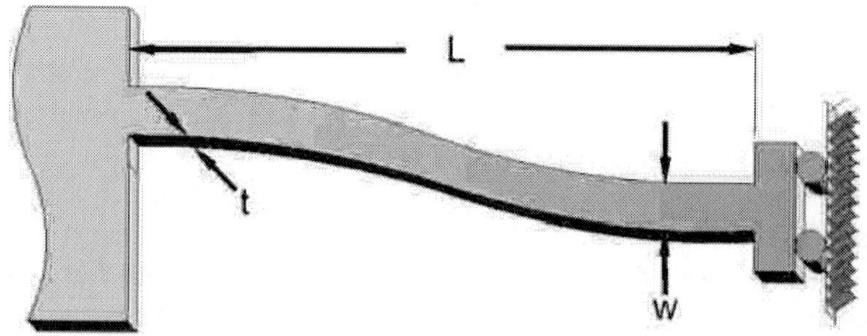

Figure 2.2. Guided-end cantilever beam.

The translational stiffness in the orthogonal direction to the beam axis is given as:

$$k = \frac{Etw^3}{L^3} \tag{2.1}$$

where E is the Young's Modulus. The beam length, thickness, and width are: L , t, and w, respectively. The spring constants are determined under the assumption that the axial strains in the other beams are negligible.

The major damping mechanism in the gyroscope structure is the viscous effects of the air surrounding the vibratory structure, confining between the proof mass surfaces and the stationary surfaces. The damping of the structural material is usually orders of magnitude lower than the viscous damping, and is generally neglected. The resulting damping in the gyroscope is dominated by the internal friction of the air between the proof-mass and the substrate; and between the comb-drive and sense capacitor fingers. These viscous damping effects can be captured by using two general damping models: Coquette flow damping and squeeze film damping shown in Figures 2.3-2.4 [83].

Coquette flow damping occurs when two plates of an area A, separated by a distance y_0 , slide parallel to each other. The Coquette flow damping coefficient can be approximated as:

$$C_{Couette} = \mu_p p \frac{A}{y_0} \tag{2.2}$$

where μ_p is the viscosity constant for air, p is the air pressure, A is the overlap area of the parallel plate, and d is the plate distance.

Squeeze film damping occurs when two parallel plates approach each other and squeeze the fluid film that is inside. Squeeze film damping effects are more complicated; they can exhibit both damping and stiffness effects, depending on the compressibility of the fluid. Using the Hagen-Poiseuille law, squeeze film damping can be modeled as:

$$C_{Squeeze} = \mu_p p \frac{7Az_0^{\ 2}}{y_0^{\ 3}} \tag{2.3}$$

where z_0 is the width of the plate.

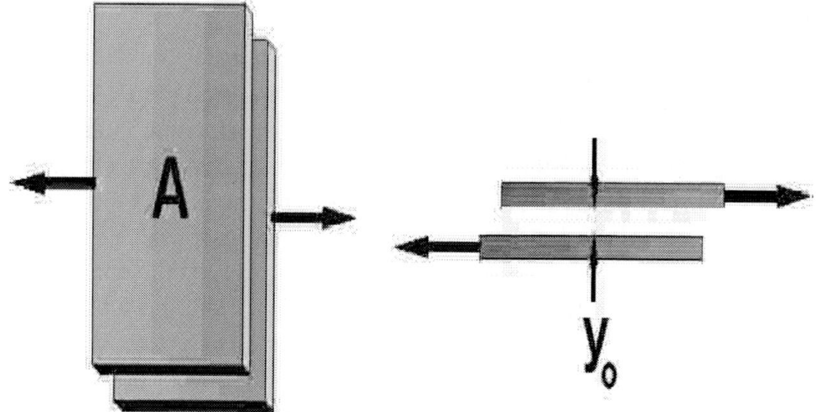

Figure 2.3. Illustration of Coquette flow damping between two plates.

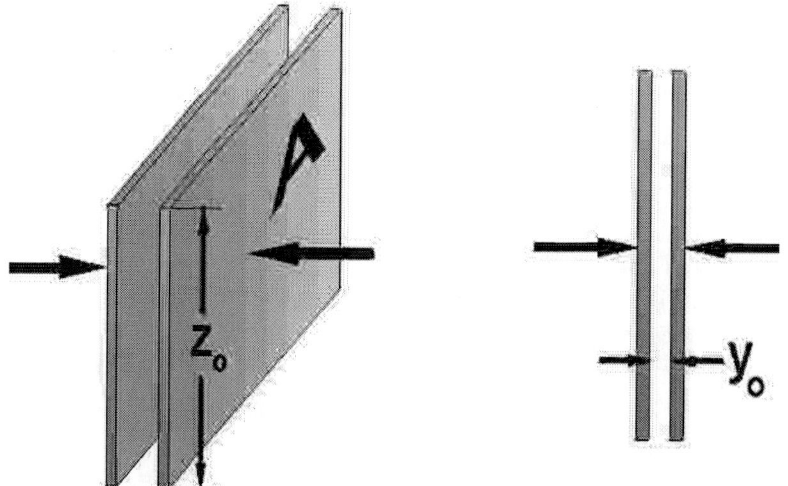

Figure 2.4. Illustration of Squeeze-film damping between two plates.

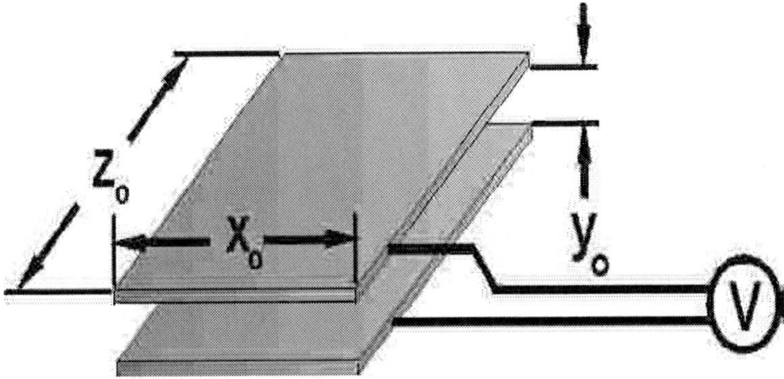

Figure 2.5. The capacitance between two plates.

Figure 2.6. Bulk-micromachining implementation of a parallel-plate actuator.

2.3. Electrical Design

In the micro-domain, capacitive sensing and actuation offer several benefits: their ease of fabrication and integration, good DC response and noise performance, high sensitivity, low drift, and low temperature sensitivity; as compared to other sensing and actuation means (piezoresistive, piezoelectric optical, magnetic, etc.).

The electrostatic actuation and sensing components of micromachined devices can be modeled as a combination of moving parallel-plate capacitors, as shown in Figure 2.5 [83]. In the most general case scenario, the capacitance between two parallel plates is expressed as:

$$C = \varepsilon_0 \varepsilon \frac{A}{y_0} = \varepsilon_0 \varepsilon \frac{x_0 z_0}{y_0} \tag{2.4}$$

where $\varepsilon_0 = 8.854 \times 10^{12}$ F/m, ε is the dielectric constant, $A = x_0 z_0$ is the total overlap area, and y_0 is the electrode gap.

In parallel-plate electrodes, an electrostatic force is generated, due to the electrostatic conservative force field between the plates. Thus, the force can be expressed as the gradient of the potential energy stored on the capacitor. Considering a voltage V applied, the electrostatic force generated by the capacitor is given by

$$F = \frac{1}{2} \frac{\partial C}{\partial y_0} V^2 = -\frac{1}{2} \frac{\varepsilon_0 \varepsilon x_0 z_0}{y_0^{\ 2}} V^2 . \tag{2.5}$$

The bulk-micromachining implementation of a parallel-plate actuator is shown in Figure 2.6 [83].

2.4. Dynamics of MEMS Gyroscope

This section discusses the dynamics of the MEMS gyroscope. A *z*-axis MEMS gyroscope is depicted in Figure 2.8. The typical MEMS vibratory gyroscope includes a proof mass, suspended by a spring, and an electrostatic actuation and sensing mechanism, for forcing an oscillatory motion and sensing the position of the proof mass. The proof mass is suspended with the help of four springs that are fixed to the gyroscope table. As such, the proof mass is free to move in *x*, *y* directions, also parallel and perpendicular to the gyroscope table, respectively. In this design, the external force is the electrostatic force. A Coriolis force acting in the *y* direction is generated whenever the proof mass undergoes rotation about the *z* axis.

The inertial frame and rotating frame are shown in Figure 2.7.

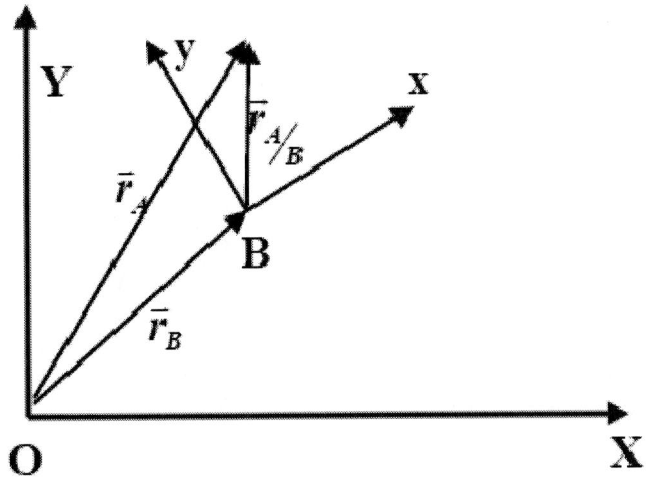

Figure 2.7. Inertial frame and rotating frame.

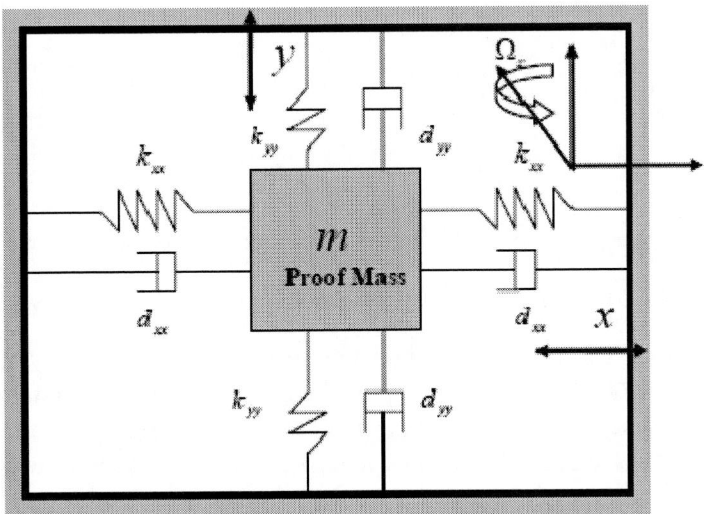

Figure 2.8. A simple model of a MEMS Z axis gyroscope.

In the Figure 2.7, $\boldsymbol{r}_A$ is a position vector of an arbitrary point A , as measured against the axes of the inertial frame. $\boldsymbol{r}_{A/B}$ is an arbitrary point A ,relative to the origin of the rotating axes ; $\boldsymbol{r}_B$ is position vector of the origin of the rotating frame, relative to the origin of the inertial frame B . Their relationship can be expressed as:

$$\begin{aligned} \boldsymbol{r}_A &= \boldsymbol{r}_B + \boldsymbol{r}_{A/B} \\ &= \boldsymbol{r}_B + x i + y j + z k. \end{aligned} \tag{2.6}$$

The velocity vector of an arbitrary point A , as measured against the axes of the inertial frame can be derived as:

$$\boldsymbol{v}_A = \boldsymbol{v}_B + \frac{dx}{dt} i + \frac{dy}{dt} j + \frac{dz}{dt} k + x\frac{di}{dt} + y\frac{dj}{dt} + z\frac{dk}{dt} . \tag{2.7}$$

Since $\frac{di}{dt} = \Omega_z \times i$, $\frac{dj}{dt} = \Omega_z \times j$, $\frac{dk}{dt} = \Omega_z \times k$, substituting these properties into (2.7) yields:

$$\begin{aligned} \boldsymbol{v}_A &= \boldsymbol{v}_B + \frac{dx}{dt} i + \frac{dy}{dt} j + \frac{dz}{dt} k + x(\Omega_z \times i) + y(\Omega_z \times j) + k(\Omega_z \times k) \\ &= \boldsymbol{v}_B + v_{A/B} + \Omega_z \times \boldsymbol{r}_{A/B} \end{aligned} \tag{2.8}$$

where $\boldsymbol{v}_A$ and $\boldsymbol{v}_B$ are the velocities with respect to the inertial frame, and $\boldsymbol{v}_{A/B} = \frac{dx}{dt} i + \frac{dy}{dt} j + \frac{dz}{dt} k$ is the relative velocity vector of an arbitrary point A , as measured against the axes of the rotating system.

The velocity of A relative to B is, therefore, made up of two terms: the velocity measured against the rotating axes; and a component that results from the rotation of the axes, and is thus invisible to the observer in the rotating frame.

The property $\frac{d(\Omega_z \times \boldsymbol{r}_{A/B})}{dt} = \frac{d\Omega_z}{dt} \times \boldsymbol{r}_{A/B} + \Omega_z \times \boldsymbol{v}_{A/B} + \Omega_z \times \Omega_z \times \boldsymbol{r}_{A/B}$ can be proved as:

$$\frac{d(\Omega_z \times \boldsymbol{r}_{A/B})}{dt} = \frac{d\Omega_z}{dt} \times \boldsymbol{r}_{A/B} + \Omega_z \times (\boldsymbol{v}_{A/B} + \Omega_z \times \boldsymbol{r}_{A/B})$$
$$= \frac{d\Omega_z}{dt} \times \boldsymbol{r}_{A/B} + \Omega_z \times \boldsymbol{v}_{A/B} + \Omega_z \times \Omega_z \times \boldsymbol{r}_{A/B}. \quad (2.9)$$

The property $\frac{d\boldsymbol{v}_{A/B}}{dt} = a_{A/B} + \Omega_z \times \boldsymbol{v}_{A/B}$ can be proved as:

$$\frac{d\boldsymbol{v}_{A/B}}{dt} = \frac{d\left(\frac{dx}{dt}i + \frac{dy}{dt}j + \frac{dz}{dt}k\right)}{dt}$$
$$= \frac{d^2x}{dt^2}i + \frac{d^2y}{dt^2}j + \frac{d^2z}{dt^2}k + \frac{dx}{dt}\cdot\frac{di}{dt} + \frac{dy}{dt}\cdot\frac{dj}{dt} + \frac{dz}{dt}\cdot\frac{dk}{dt}$$
$$= \frac{d^2x}{dt^2}i + \frac{d^2y}{dt^2}j + \frac{d^2z}{dt^2}k + \frac{dx}{dt}\cdot(\Omega_z \times i) + \frac{dy}{dt}\cdot(\Omega_z \times j) + \frac{dz}{dt}\cdot(\Omega_z \times k) \quad (2.10)$$
$$= \frac{d^2x}{dt^2}i + \frac{d^2y}{dt^2}j + \frac{d^2z}{dt^2}k + \Omega_z \times \frac{dx}{dt}i + \Omega_z \times \frac{dy}{dt}j + \Omega_z \times \frac{dz}{dt}k$$
$$= \boldsymbol{a}_{A/B} + \Omega_z \times \boldsymbol{v}_{A/B}.$$

Differentiating (2.10), using (2.8) and (2.9) yields:

$$\boldsymbol{a}_A = \boldsymbol{a}_B + \boldsymbol{a}_{A/B} + 2\Omega_z \times \boldsymbol{v}_{A/B} + \frac{d\Omega_z}{dt} \times \boldsymbol{r}_{A/B} + \Omega_z \times (\Omega_z \times \boldsymbol{r}_{A/B}) \quad (2.11)$$

where $\boldsymbol{a}_A$ and $\boldsymbol{a}_B$ are the accelerations, with respect to the inertial frame, $\boldsymbol{a}_{A/B} = \frac{d^2x}{dt^2}i + \frac{d^2y}{dt^2}j + \frac{d^2z}{dt^2}k$ is the acceleration vector of an arbitrary point A, as measured against the axes of the rotating system.

The acceleration of A relative to B is, therefore, made up of four terms: the acceleration measured against the rotating axes; and three components, resulting from the rotation of the axes, and are invisible to the observer in the rotating frame. Respectively, these are the Euler (tangential), Centripetal and Coriolis accelerations.

Multiplying (2.11) by mass m gives:

$$ma_A = ma_B + ma_{A/B} + 2m\,\Omega_z \times v_{A/B} + m\left(\frac{d\Omega_z}{dt} \times r_{A/B}\right) + m\Omega_z \times (\Omega_z \times r_{A/B}) \quad (2.12)$$

where $2m\Omega_z \times v_{A/B}$ is the Coriolis force, and $m\Omega_z \times (\Omega_z \times r_{A/B})$ is the Centrifugal force.

The Coriolis force, acting on the proof mass along x direction, is derived as :

$$F_{coriolis-x} = 2m\Omega_z k \times \dot{y}j = 2m\Omega_z \dot{y}i\,. \quad (2.13)$$

The Coriolis force, acting on the proof mass along y direction, is derived as:

$$F_{coriolis-y} = 2m\Omega_z k \times \dot{x}i = 2m\Omega_z \dot{x}j\,. \quad (2.14)$$

By using the property of $k \times (k \times i) = -i$; the Centripetal force, acting on the proof mass along x direction, can be derived as:

$$F_{centripetal-x} = m\Omega_z k \times (\Omega_z k \times xi) = -m\Omega_z^2 xi\,. \quad (2.15)$$

By using the property of $k \times (k \times j) = -j$; the Centripetal force, acting on the proof mass along y direction, can be derived as:

$$F_{centripetal-y} = m\Omega_z k \times (\Omega_z k \times yj) = -m\Omega_z^2 yj\,. \quad (2.16)$$

We assume that the table where the proof mass is mounted, is moving with a constant velocity; the gyroscope is rotating at a constant angular velocity Ω_z over a sufficiently long time interval; the centripetal forces, $m\Omega_z^2 x$ and $m\Omega_z^2 y$, are assumed to be negligible; and the gyroscope undergoes a rotation about the z axis only. Therefore, the Coriolis force is generated in a direction perpendicular to the drive and rotational axes.

With these assumptions, the dynamics of the gyroscope now becomes:

$$m\ddot{x} + d_{xx}\dot{x} + d_{xy}\dot{y} + k_{xx}x + k_{xy}y = u_x + 2m\Omega_z \dot{y} \quad (2.17)$$

$$m\ddot{y} + d_{xy}\dot{x} + d_{yy}\dot{y} + k_{xy}x + k_{yy}y = u_y - 2m\Omega_z\dot{x}. \tag{2.18}$$

Fabrication imperfections contribute mainly to the symmetric spring and damping terms, k_{xy} and d_{xy}. The *x* and *y* axes spring and damping terms: k_{xx}, k_{yy}, d_{xx} and d_{yy}, are mostly known, but have small unknown variations from their nominal values. The mass of the proof mass can be determined with great accuracy, and u_x, u_y are the control forces in the x and y direction.

Dividing (2.17) and (2.18) by the reference mass and rewriting the gyroscope dynamics in vector forms, results in:

$$\ddot{\boldsymbol{q}} + \frac{\boldsymbol{D}}{m}\dot{\boldsymbol{q}} + \frac{\boldsymbol{K}_a}{m}\boldsymbol{q} = \frac{\boldsymbol{u}}{m} - 2\boldsymbol{\Omega}\dot{\boldsymbol{q}} \tag{2.19}$$

where $\boldsymbol{q} = \begin{bmatrix} x \\ y \end{bmatrix}$, $\boldsymbol{u} = \begin{bmatrix} u_x \\ u_y \end{bmatrix}$, $\boldsymbol{\Omega} = \begin{bmatrix} 0 & -\Omega_z \\ \Omega_z & 0 \end{bmatrix}$, $\boldsymbol{D} = \begin{bmatrix} d_{xx} & d_{xy} \\ d_{xy} & d_{yy} \end{bmatrix}$,

$\boldsymbol{K}_a = \begin{bmatrix} k_{xx} & k_{xy} \\ k_{xy} & k_{yy} \end{bmatrix}$.

Using non-dimensional time $t^* = w_0 t$; dividing both sides of (2.19) by w_0^2 and reference length q_0, gives the final form of the non-dimensional equation of motion as:

$$\frac{\ddot{\boldsymbol{q}}}{q_0} + \frac{\boldsymbol{D}}{mw_0}\frac{\dot{\boldsymbol{q}}}{q_0} + \frac{\boldsymbol{K}}{mw_0^2}\frac{\boldsymbol{q}}{q_0} = \frac{\boldsymbol{u}}{mw_0^2 q_0} - 2\frac{\boldsymbol{\Omega}}{w_0}\frac{\dot{\boldsymbol{q}}}{q_0}. \tag{2.20}$$

Defining a set of new parameters as follows:

$$\boldsymbol{q}^* = \frac{\boldsymbol{q}}{q_0}, \boldsymbol{D}^* = \frac{\boldsymbol{D}}{mw_0}, \Omega_z{}^* = \frac{\Omega_z}{w_0}, \tag{2.21}$$

$$\boldsymbol{u}^* = \frac{\boldsymbol{u}}{mw_0^{\,2} q_0}, w_x = \sqrt{\frac{k_{xx}}{mw_0^{\,2}}}, w_y = \sqrt{\frac{k_{yy}}{mw_0^{\,2}}}, w_{xy} = \frac{k_{xy}}{mw_0^{\,2}}. \quad (2.22)$$

Ignoring the superscript (*) for notational clarity, the nondimensional representation is:

$$\ddot{\boldsymbol{q}} + \boldsymbol{D}\dot{\boldsymbol{q}} + \boldsymbol{K}_b \boldsymbol{q} = \boldsymbol{u} - 2\boldsymbol{\Omega}\dot{\boldsymbol{q}} \quad (2.23)$$

where $\boldsymbol{K}_b = \begin{bmatrix} w_x^{\,2} & w_{xy} \\ w_{xy} & w_y^{\,2} \end{bmatrix}$.

2.5. Dynamics of MEMS Triaxial Gyroscope

Assume: the gyroscope is moving with a constant linear speed, the gyroscope is rotating at a constant angular velocity, the centrifugal forces are assumed negligible, and the gyroscope undergoes rotations along the x, y and z axes. Referring to [58], the dynamics equations of the triaxial gyroscope system are as follows:

$$m\ddot{x} + d_{xx}\dot{x} + d_{xy}\dot{y} + d_{xz}\dot{z} + k_{xx}x + k_{xy}y + k_{xz}z = u_x + 2m\Omega_z\dot{y} - 2m\Omega_y\dot{z}$$

$$m\ddot{y} + d_{xy}\dot{x} + d_{yy}\dot{y} + d_{yz}\dot{z} + k_{xy}x + k_{yy}y + k_{yz}z = u_y - 2m\Omega_z\dot{x} + 2m\Omega_x\dot{z} \quad (2.24)$$

$$m\ddot{z} + d_{xz}\dot{x} + d_{yz}\dot{y} + d_{zz}\dot{z} + k_{xz}x + k_{yz}y + k_{zz}z = u_z + 2m\Omega_y\dot{x} - 2m\Omega_x\dot{y}$$

where m is the mass of the proof mass and Fabrication imperfections contribute mainly to the asymmetric spring terms: w_{xy}, w_{xz} and w_{yz}; and asymmetric damping terms: d_{xy}, d_{yz} and d_{xz}; w_x, w_y and w_z are spring terms in the x, y and z directions, respectively; d_{xx}, d_{yy} and d_{zz} are damping terms in the x, y and z directions, respectively; Ω_x, Ω_y and Ω_z are angular velocities in the x, y and z directions, respectively; u_x, u_y and u_z are the control forces in the x, y and z directions, respectively.

Dividing the equation by the reference mass and rewriting the dynamics in vector forms, results in:

$$\ddot{q}+\frac{D}{m}\dot{q}+\frac{K_a}{m}q=\frac{u}{m}-2\Omega\dot{q} \tag{2.25}$$

where $q=\begin{bmatrix} x \\ y \\ z \end{bmatrix}$, $u=\begin{bmatrix} u_x \\ u_y \\ u_z \end{bmatrix}$, $\Omega=\begin{bmatrix} 0 & -\Omega_z & \Omega_y \\ \Omega_z & 0 & -\Omega_x \\ -\Omega_y & \Omega_x & 0 \end{bmatrix}$,

$$D=\begin{bmatrix} d_{xx} & d_{xy} & d_{xz} \\ d_{xy} & d_{yy} & d_{yz} \\ d_{xz} & d_{yz} & d_{zz} \end{bmatrix},\ K_a=\begin{bmatrix} k_{xx} & k_{xy} & k_{xz} \\ k_{xy} & k_{yy} & k_{yz} \\ k_{xz} & k_{yz} & k_{zz} \end{bmatrix}.$$

Because of the non-dimensional time, $t^*=w_0 t$, dividing both sides of equation by reference frequency w_0^2 and reference length q_0, gives the final form of the non-dimensional equation of motion for the z-axis gyroscope.

$$\frac{\ddot{q}}{q_0}+\frac{D}{mw_0}\frac{\dot{q}}{q_0}+\frac{K_a}{mw_0^2}\frac{q}{q_0}=\frac{u}{mw_0^2 q_0}-2\frac{\Omega}{w_0}\frac{\dot{q}}{q_0}. \tag{2.26}$$

Define new parameters as follows:

$$q^*=\frac{q}{q_0},\ D^*=\frac{D}{mw_0},\ \Omega^*=\frac{\Omega}{w_0},$$

$$u_x^*=\frac{u_x}{mw_0^2 q_0},\ u_y^*=\frac{u_y}{mw_0^2 q_0},\ u_z^*=\frac{u_z}{mw_0^2 q_0},$$

$$w_x=\sqrt{\frac{k_{xx}}{mw_0^2}},\ w_y=\sqrt{\frac{k_{yy}}{mw_0^2}},\ w_z=\sqrt{\frac{k_{zz}}{mw_0^2}},$$

$$w_{xy} = \frac{k_{xy}}{mw_0^2},\ w_{yz} = \frac{k_{yz}}{mw_0^2},\ w_{xz} = \frac{k_{xz}}{mw_0^2}.$$

Ignoring the superscript for notational clarity, the non-dimensional representation is:

$$\ddot{q} + D\dot{q} + K_b q = u - 2\Omega\dot{q} \tag{2.27}$$

where $K_b = \begin{bmatrix} w_x^2 & w_{xy} & w_{xz} \\ w_{xy} & w_y^2 & w_{yz} \\ w_{xz} & w_{yz} & w_z^2 \end{bmatrix}$.

2.6. State-Space Representation of MEMS Gyroscope

Rewriting the gyroscope model (2.23) in state space form as:

$$\dot{\boldsymbol{X}} = \boldsymbol{A}\boldsymbol{X} + \boldsymbol{B}\boldsymbol{u} \tag{2.28}$$

where

$$\boldsymbol{A} = \begin{bmatrix} 0 & 1 & 0 & 0 \\ -w_x^2 & -d_{xx} & -w_{xy} & -(d_{xy} - 2\Omega_z) \\ 0 & 0 & 0 & 1 \\ -w_{xy} & -(d_{xy} + 2\Omega_z) & -w_y^2 & -d_{yy} \end{bmatrix}$$

$$\boldsymbol{B} = \begin{bmatrix} 0 & 1 & 0 & 0 \\ 0 & 0 & 0 & 1 \end{bmatrix}^T,\quad \boldsymbol{u} = \begin{bmatrix} u_x \\ u_y \end{bmatrix},\quad \boldsymbol{X} = \begin{bmatrix} x \\ \dot{x} \\ y \\ \dot{y} \end{bmatrix}.$$

The reference model, $x_m = A_1 \, sin(w_1 t)$, $y_m = A_2 \, sin(w_2 t)$; is defined as:

$$\ddot{\boldsymbol{q}}_m + \boldsymbol{K}_m \boldsymbol{q}_m = 0 \tag{2.29}$$

where $\boldsymbol{K}_m = diag\{w_1^{\ 2} \quad w_2^{\ 2}\}$.

Similar to (2.28), the reference model can be written as:

$$\dot{\boldsymbol{X}}_m = \begin{bmatrix} 0 & 1 & 0 & 0 \\ -w_1^{\ 2} & 0 & 0 & 0 \\ 0 & 0 & 0 & 1 \\ 0 & 0 & -w_2^{\ 2} & 0 \end{bmatrix} \boldsymbol{X}_m \equiv \boldsymbol{A}_m \boldsymbol{X}_m \tag{2.30}$$

where $\boldsymbol{A}_m$ is a known constant matrix.

Chapter 3

SYSTEM IDENTIFICATION OF MEMS GYROSCOPES

Fabrication defects and perturbations affect the behavior of a vibratory MEMS gyroscope sensor, making it difficult to measure the rotation angular rate. This chapter presents a novel adaptive approach that can identify, in an on-line fashion, angular rate and other system parameters. The proposed approach develops a series-parallel on-line identifier by rewriting the dynamic model of a MEMS gyroscope sensor, updating the estimator of angular rate adaptively. The feasibility of the proposed approach is analyzed and proved by Lyapunov's direct method. These simulation results reveal the validity and effectiveness of the on-line identifier.

3.1. THE DESIGN OF AN ON-LINE IDENTIFIER

In this chapter, a novel adaptive online identifier is designed to estimate the angular rate and system parameters. The motivation of this chapter is to propose a novel series-parallel on-line identifier that could estimate all the system parameters using observer state and control signals. The advantage of the proposed adaptive approach is that it is easy to implement in practice and it avoids the complicated algorithm derivation.. To some extent, therefore, it is better than other control algorithms for the vibratory MEMS gyroscope. In (2.28), this chapter will develop a novel adaptive approach that can identify the system matrix A in an on-line fashion.

The overall objective of this section is to generate an adaptive law for identifying $\boldsymbol{A}$ on-line, by using the observed signals $\mathrm{x}(t)$ and $\mathrm{u}(t)$. Figure 3.1 shows the block diagram of the on-line identifier.

To start, we rewrite the dynamic model (2.28) by adding and subtracting a term, $\boldsymbol{A}_m\mathrm{x}$; where $\boldsymbol{A}_m$ is an arbitrary stable matrix:

$$\dot{\mathrm{x}} = \boldsymbol{A}_m\mathrm{x} + (\boldsymbol{A} - \boldsymbol{A}_m)\mathrm{x} + \boldsymbol{B}\mathrm{u} \tag{3.1}$$

The adaptive law for generating the estimate, $\hat{\boldsymbol{A}}$ of $\boldsymbol{A}$, is to be driven by the estimation error:

$$\varepsilon = \mathrm{x} - \hat{\mathrm{x}} \tag{3.2}$$

where $\hat{\mathrm{x}}$ is the estimated value of x, using the estimate $\hat{\boldsymbol{A}}$. The state of $\hat{\mathrm{x}}$ is generated by an equation that has the same form as the gyroscope sensor model, but with $\boldsymbol{A}$ replaced by $\hat{\boldsymbol{A}}$. Considering the gyroscope sensor equation (3.1), this chapter generates $\hat{\mathrm{x}}$ by a series-parallel model:

$$\dot{\hat{\mathrm{x}}} = \boldsymbol{A}_m\hat{\mathrm{x}} + (\hat{\boldsymbol{A}} - \boldsymbol{A}_m)\mathrm{x} + \boldsymbol{B}\mathrm{u} \tag{3.3}$$

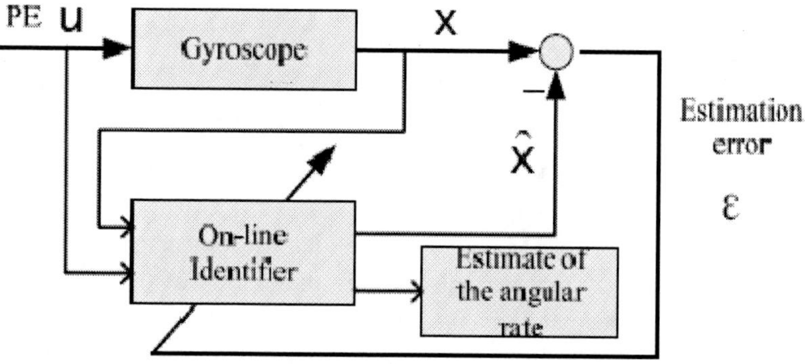

Figure 3.1. Block diagram of the on-line identifier.

The estimation error, $\varepsilon = \mathrm{x} - \hat{\mathrm{x}}$, satisfies the differential equation:

$$\dot{\varepsilon} = \boldsymbol{A}_m\varepsilon - \tilde{\boldsymbol{A}}\mathrm{x} \tag{3.4}$$

where $\tilde{A} = \hat{A} - A$.

Equation (3.4) indicates how the parameter error affects the estimation errors ε. Because A_m is stable, a zero parameter error implies that ε converges to zero, exponentially. Indeed, $\tilde{A}$ is unknown and ε is the only measured signal that we can monitor in practice to check the accuracy of the estimation. Utilizing the measured signals, we assume the adaptive law is of the form:

$$\dot{\hat{A}} = \mathrm{F}(\varepsilon, \mathrm{x}, \hat{\mathrm{x}}, \mathrm{u}) \tag{3.5}$$

where, F is the function of measured signals and is to be chosen so that the equilibrium state is:

$$\hat{A}_e = A, \quad \varepsilon_e = 0 \tag{3.6}$$

of the differential equation described by (3.4) and (3.6) is uniformly stable; or if possible, uniformly asymptotically stable; or, even better, exponentially stable.

Considering the following Lyapunov function candidate:

$$\mathrm{V}(\varepsilon, \tilde{A}) = \varepsilon^T P \varepsilon + \mathrm{tr}(\tilde{A}^T \tilde{A}) \tag{3.7}$$

where $\mathrm{tr}(A)$ denotes the trace of a matrix A, and $P = P^T > 0$ is chosen as the solution of the Lyapunov equation:

$$A_m^{\ T} P + P A_m = -Q \tag{3.8}$$

where $Q = Q^T > 0$, whose existence is guaranteed by the stability of A_m.

The time derivative, $\dot{\mathrm{V}}$ of V, along the trajectory of equation (3.4),(3.7) is:

$$\begin{aligned} \dot{\mathrm{V}} &= \varepsilon^T P \dot{\varepsilon} + \dot{\varepsilon}^T P \varepsilon + \mathrm{tr}(\dot{\tilde{A}}^T \tilde{A} + \tilde{A}^T \dot{\tilde{A}}) \\ &= \varepsilon^T (A_m^{\ T} P + P A_m) \varepsilon - 2\varepsilon^T P \tilde{A} \mathrm{x} + tr(\dot{\tilde{A}}^T \tilde{A} + \tilde{A}^T \dot{\tilde{A}}) \end{aligned} \tag{3.9}$$

Using the properties of trace of a matrix:

$$\dot{V} = -\varepsilon^T \boldsymbol{Q}\varepsilon + 2\,\mathrm{tr}(\dot{\tilde{\boldsymbol{A}}}\tilde{\boldsymbol{A}}^T - \boldsymbol{P}\varepsilon\, x^T \tilde{\boldsymbol{A}}^T) \tag{3.10}$$

The obvious choice for $\dot{\tilde{\boldsymbol{A}}}$ to make $\dot{V}$ negative is:

$$\dot{\tilde{\boldsymbol{A}}} = \dot{\hat{\boldsymbol{A}}} = F = \boldsymbol{P}\varepsilon\, x^T \ . \tag{3.11}$$

This adaptive law yields:

$$\dot{V} = -\varepsilon^T \boldsymbol{Q}\varepsilon \le 0 \ . \tag{3.12}$$

Now, let's consider the boundedness of the state vector x. The dynamics of the MEMS gyroscope sensor, as described by the equations (2.17-2.18), may be considered as a mass: a spring and damper system, implying that $\boldsymbol{A}$ is stable. With the assumption that the control inputs, u_x and u_y , are bounded, x is also ensured to be bounded.

Equation (3.7) implies that the equilibrium $\hat{\boldsymbol{A}}_e = \boldsymbol{A}$, $\varepsilon_e = 0$ of the respective equations is uniformly stable and $\tilde{\boldsymbol{A}}$, ε are bounded. Using equation (3.4), $\dot{\varepsilon}$ is bounded. Since:

$$\dot{V} = -\varepsilon^T \boldsymbol{Q}\varepsilon \le -\lambda_{\min}(\boldsymbol{Q})\left|\varepsilon\right|^2 \le 0 \tag{3.13}$$

where $\lambda_{\min}(\boldsymbol{Q})$ is the minimum eigenvalue of $\boldsymbol{Q}$ and satisfies $\lambda_{\min} > 0$. The inequality (3.13) implies that ε is integrable as $\int_0^t \left|\varepsilon\right|^2 dt \le 1/\lambda_{\min}\,[V(0) - V(t)]$. Since $V(0)$ is bounded and $0 \le V(t) \le V(0)$, it can be concluded that $\lim_{t\to\infty}\int_0^t \left|\varepsilon\right|^2 dt$ is bounded. Since $\lim_{t\to\infty}\int_0^t \left|\varepsilon\right|^2 dt$, ε and $\dot{\varepsilon}$ are bounded, according to Barbalat's lemma, $\lim_{t\to\infty}\varepsilon(t) = 0$; in turn, implying that $\left\|\dot{\hat{A}}\right\| \to \boldsymbol{0}$.

Like in other adaptive control problems, the persistent excitation condition is an important factor in correctly estimating the angular rate. From (2.28) the dynamics of a MEMS gyroscope sensor can be considered as a fourth-order system, implying that if the control input u contains two different non-zero frequencies, the persistency of excitation is satisfied. Then, we define:

$$u_x = A_1 \sin(w_1 t)\,,\quad u_y = A_2 \sin(w_2 t) \tag{3.14}$$

where w_1, w_2 satisfies $w_1 \neq w_2$, $w_1 \neq 0$, $w_2 \neq 0$. Under these assumptions, the estimate $\hat{\boldsymbol{A}}$ converges to its true value $\boldsymbol{A}$.

In summary, if $u_x = A_1\, sin(\, w_1 t\,)$ and $u_y = A_2\, sin(\, w_2 t\,)$ are used, then ε converges to zero asymptotically. Consequently, the angular rate and system parameters converge to their true values.

3.2. Simulation Study

In this section, we will evaluate the proposed adaptive approach on the lumped MEMS gyroscope sensor model. The objective of the adaptive approach is to identify the parameters correctly, including the input angular velocity. Parameters of the MEMS gyroscope sensor are as follows:

$m = 1.8\times10^{-7}\,kg$, $k_{xx} = 63.955N/m$, $k_{yy} = 95.92N/m$, $k_{xy} = 12.779N/m$, $d_{xx} = 1.8\times10^{-6}\,Ns/m$, $d_{yy} = 1.8\times10^{-6}\,Ns/m$, $d_{xy} = 3.6\times10^{-7}\,Ns/m$

Since the general displacement range of the MEMS gyroscope sensor in each axis is in the sub-micrometer level, it is reasonable to choose 1 μm as the reference length q_0. Given that the usual natural frequency for each of the axles of a vibratory MEMS gyroscope sensor is in the KHz range, the w_0 as 1KHz is chosen. The unknown angular velocity is assumed as $\Omega_z = 100 rad/s$. Consequently, the non-dimensional values of the MEMS gyroscope sensor parameters are as follows:

$w_x^{\,2} = 355.3$, $w_y^{\,2} = 532.9$, $w_{xy} = 70.99$, $d_{xx} = 0.01$, $d_{yy} = 0.01$, $d_{xy} = 0.002$, $\Omega_z = 0.1$.

Given the arbitrariness of stable matrix, $\boldsymbol{A}_m$, we choose $\boldsymbol{A}_m = -20*\boldsymbol{I}$ for the simplicity in choosing the gain matrix $\boldsymbol{P}$. With $\boldsymbol{A}_m = -20*\boldsymbol{I}$, $\boldsymbol{P}$ can be an arbitrary positive symmetrical matrix. It should be noted that the choosing of $\boldsymbol{P}$ should take into account of the balance of the elements in $\dot{\hat{\boldsymbol{A}}}$. The control input forces are $u_x = 400\sin(2t)$, $u_y = 400\sin(10t)$, and the initial value of $\hat{A}(0) = 0.9 * \boldsymbol{A}$. The simulation results are shown in Figure 3,4,5.

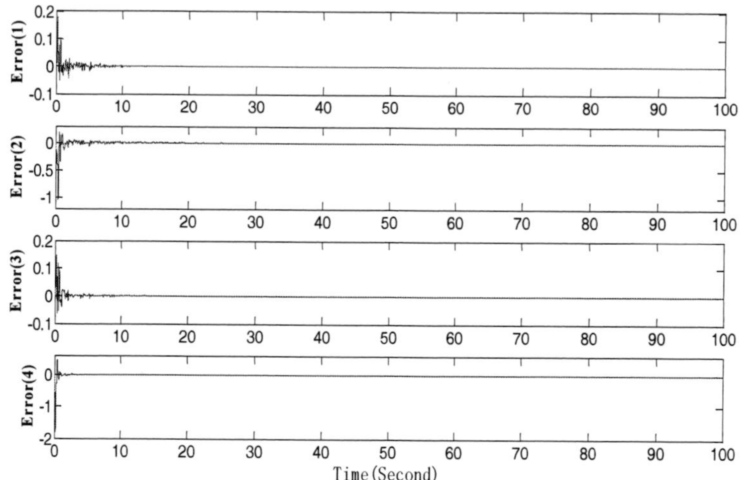

Figure 3.2. The estimation errors.

Figure 3.2 depicts the estimation errors. It is observed that the estimation errors converge to zero very quickly, under the sinusoidal input of two different non-zero frequencies. This validates that the equilibrium $\varepsilon_e = 0$ is uniformly asymptotically stable. Figure 3.3 clearly shows that the estimation of the angular rate reaches its actual value in finite time. The regulating time is about 10 seconds and the overshot is approx 49%. Simulation experiments show that the regulating time and the overshot of the angular rate estimation are a pair of contradictions. Choosing a reasonable matrix $\boldsymbol{P}$ could prove to find a compromise between them. Figure 3.4 shows that all of the estimated values of w_x^2, w_y^2, w_{xy} can reach their true values in a relatively very short time and with very small overshot.

The simulation results verify that with the series-parallel identifier model (3.1) and the parameter adaptive law (3.11); if the persistent excitation condition is satisfied, that is, two different non-zero frequencies inputs; the estimation errors converge to zero asymptotically, and all unknown parameters (including the angular velocity) quickly go to their true values and without large overshot.

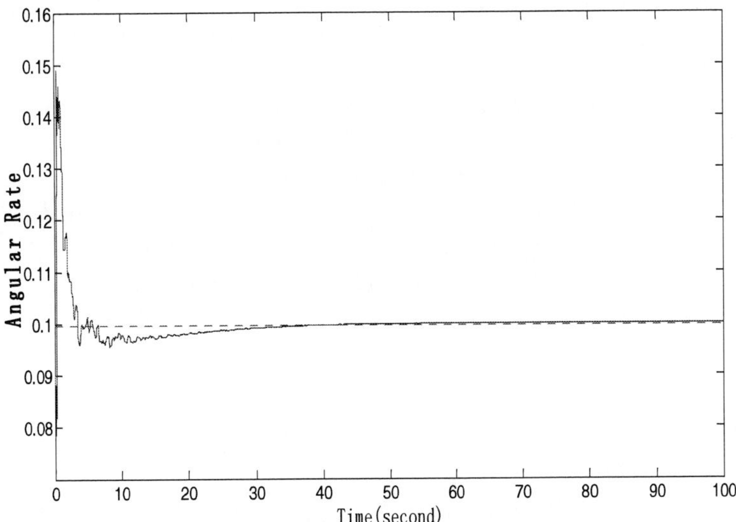

Figure 3.3. Adaptation of angular velocity.

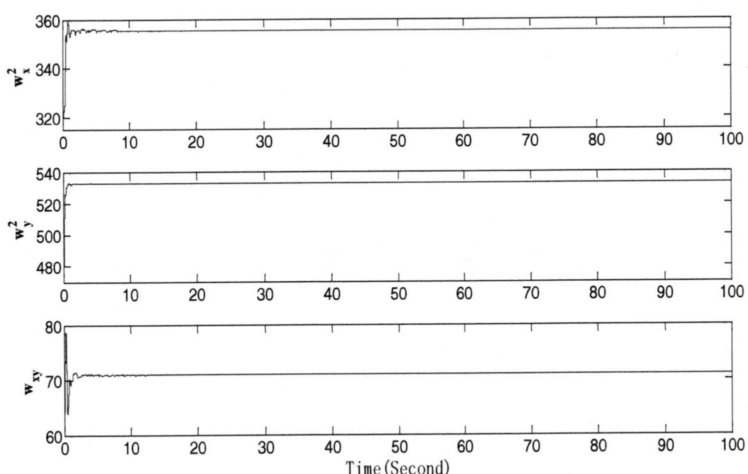

Figure 3.4. Adaptation of $w_x^{\ 2}$, $w_y^{\ 2}$, w_{xy} .

Concluding Remarks

This chapter investigates the design of adaptive control for the MEMS gyroscope sensor. The dynamics model of the MEMS gyroscope sensor is developed and nondimensionized. A novel adaptive approach with series-parallel on-line identifier is proposed, and a stability condition is established. Simulation results demonstrate the effectiveness of the proposed adaptive on-line identifier in identifying the gyroscope sensor parameters and angular rate. It should be recognized that model uncertainties and external disturbances, which should be compensated for in a real application, have not yet been considered in the proposed on-line adaptive. The next step is to incorporate the terms of model uncertainties and external disturbances into the online identifier, improving the robustness of the proposed method.

Chapter 4

ADAPTIVE CONTROL OF MEMS GYROSCOPE

This chapter presents an adaptive control approach for a MEMS z-axis gyroscope sensor. The dynamical model of a MEMS gyroscope sensor is derived and an adaptive state tracking control for a MEMS gyroscope is developed. The proposed adaptive control approaches can estimate the both angular velocity and the damping and stiffness coefficients; including the coupling terms, due to the fabrication imperfection. With the proposed adaptive control strategy, the stability of the closed-loop systems is established. Numerical simulation is investigated to verify the effectiveness of the proposed control scheme.

4.1. THE DESIGN OF ADAPTIVE CONTROL AND STABILITY ANALYSIS

In this section, to identify the angular velocity and all unknown gyroscope parameters, an adaptive controller is proposed. A block diagram is shown in Figure 4.1. In the adaptive control design, we consider the equation (2.28) as the system model.

Two assumptions are made: First, there exists a constant matrix $\boldsymbol{K}^*$, such that the matching condition $\boldsymbol{A}+\boldsymbol{B}\boldsymbol{K}^{*T}=\boldsymbol{A}_m$ can always be satisfied. Second, the control target for the MEMS gyroscope is (i) to design an adaptive controller so that the trajectory of $\boldsymbol{X}$ can track the state of reference model

X_m ; (ii) to estimate the angular velocity of the MEMS gyroscope and all unknown gyroscope parameters.

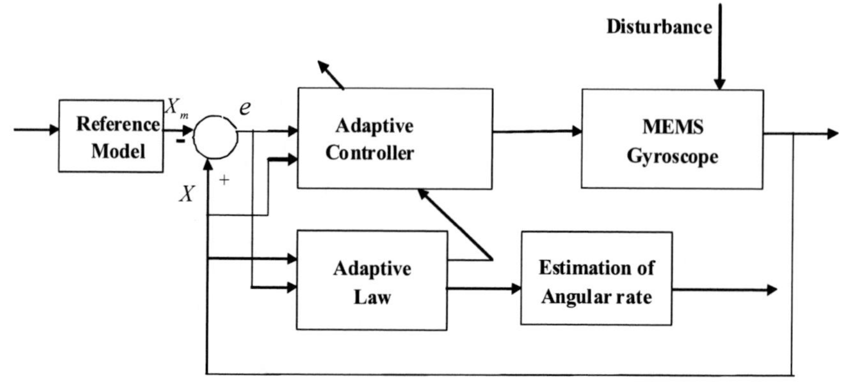

Figure 4.1. Block diagram of adaptive control for MEMS gyroscope.

The tracking error and its derivative are:

$$e(t) = X(t) - X_m(t) \tag{4.1}$$

$$\dot{e} = A_m e + (A - A_m)X + Bu\,. \tag{4.2}$$

The adaptive controller is proposed as:

$$u(t) = K^T(t)X(t) + K_f e \tag{4.3}$$

where $K(t)$ is an estimate of K^*, and the constant matrix K_f satisfies the condition that $(A_m + BK_f)$ is Hurwitz.

We define the estimation error as $\widetilde{K}(t) = K(t) - K^*$ and substitute this estimation error and (4.3) into (2.28) yield

$$\dot{X}(t) = A_m X(t) + B\widetilde{K}^T(t)X(t) \tag{4.4}$$

Then, we have the tracking error equation:

$$\dot{e}(t) = (A_m + BK_f)e + B\tilde{K}^T(t)X(t) \ . \tag{4.5}$$

Defining a Lyapunov function:

$$V = \frac{1}{2}e^T Pe + \frac{1}{2}tr\left[\tilde{K}M^{-1}\tilde{K}^T\right] \tag{4.6}$$

where P and $M = diag\{m_1 \quad m_2\}$ are a positive definite matrix.

Differentiating V with respect to time yields:

$$\begin{aligned}\dot{V} &= e^T P\dot{e} + tr\left[\tilde{K}M^{-1}\dot{\tilde{K}}^T\right] \\ &= -e^T Qe + e^T PB\tilde{K}^T X + tr\left[\tilde{K}M^{-1}\dot{\tilde{K}}^T\right]\end{aligned} \tag{4.7}$$

where $P(A_m + BK_f) + (A_m + BK_f)^T P = -Q$, Q is a positive definite matrix.

To make $\dot{V} \le 0$, we choose the adaptive law as:

$$\dot{\tilde{K}}^T(t) = \dot{K}^T(t) = -MB^T P^T eX^T(t) \tag{4.8}$$

with $K(0)$ being arbitrary. This adaptive law yields:

$$\dot{V} = -e^T Qe \le -\lambda_{min}(Q)\|e\| \le 0 \tag{4.9}$$

The inequality (4.9) implies that e is as integrable as $\int_0^t \|e\| dt \le \frac{1}{\lambda_{min}(Q)}\left[V(0) - V(t)\right]$ .Since $V(0)$ is bounded and $V(t)$ is nonincreasing and bounded, it can be concluded that $\lim_{t\to\infty}\int_0^t \|e\| dt$ is likewise bounded. Since $\lim_{t\to\infty}\int_0^t \|e\| dt$ and $\dot{e}$ are both bounded, e will asymptotically converge to zero, $\lim_{t\to\infty} e(t) = 0$, according to Barbalat's lemma.

It can be shown that if the persistent excitation can be satisfied, i.e. $w_1 \neq w_2$, the controller parameter converges to its true values, $\widetilde{\boldsymbol{K}} \to 0$. In other words, an excitation of proof mass should persistently be exciting. Since $\widetilde{\boldsymbol{K}} \to 0$, then the unknown angular velocity, as well as all other unknown parameters, can be determined from $\boldsymbol{A} + \boldsymbol{B}\boldsymbol{K}^T = \boldsymbol{A}_m$ and we obtain $\Omega_z = 0.25(k_{22} - k_{41})$.

Conclusion: if persistently exciting drive signals, $x_m = A_1 \sin(w_1 t)$ and $y_m = A_2 \sin(w_2 t)$, are used; then $\widetilde{\boldsymbol{K}}$ and $\boldsymbol{e}(t)$ all converge to zero asymptotically. Consequently, the unknown angular velocity can be determined as $\lim_{t\to\infty} \Omega_z(t) = \Omega_z$. Nonetheless, it is difficult to establish the convergence rate.

4.2. Simulation Example

Using MATLAB/SIMULINK, we will evaluate the proposed adaptive control on the lumped MEMS gyroscope model. The control objective is to design an adaptive state tracking controller, obtaining consistent estimate of Ω_z.

In the simulation, we allowed $\pm 5\%$ parameter variations for the spring and damping coefficients, with respect to their nominal values. We further assumed $\pm 5\%$ magnitude changes in the coupling terms, i.e. d_{xy} and ω_{xy}, similarly with respect to their nominal values. The external disturbance is a random variable signal, with zero mean and unity variance. The parameters of the MEMS gyroscope are as follows:

$m = 0.57e-8$ kg, $d_{xx} = 0.429e-6$ N s/m, $d_{xy} = 0.0429e-6$ N s/m, $d_{yy} = 0.687e-36$ N s/m,

$k_{xx} = 80.98$ N/m, $k_{xy} = 5$ N/m, $k_{yy} = 71.62$ N/m, $w_0 = 1kHz$, $q_0 = 10^{-6} m$.

The unknown angular velocity is assumed as $\Omega_z = 5.0$ rad/s, and the initial condition on the $\boldsymbol{K}$ matrix is $\boldsymbol{K}(0) = 0.95\boldsymbol{K}^*$. The desired motion trajectories are $x_m = \sin(w_1 t)$ and $y_m = 1.2\sin(w_2 t)$, where $w_1 = 6.17kHz$ and $w_2 = 5.11kHz$. The adaptive gain of (4.8) $\boldsymbol{M} = diag\{20 \quad 20\}$. The $\boldsymbol{K}_f$ in (4.5) is chosen as $\boldsymbol{K}_f = \begin{bmatrix} -10000 & -10000 & 1000 & 20000 \\ -1000 & -1000 & -1000 & -1000 \end{bmatrix}$.

Figure 4.2 depicts the tracking errors. It can be observed that the tracking errors converge to zero asymptotically. Figures 4.3 and 4.4 draw the adaptation of the controller parameters and angular velocity. It is illustrated that the estimates of angular velocity and controller parameters converge to their values. Using adaptive control, Figure 4.5 plots the control input.

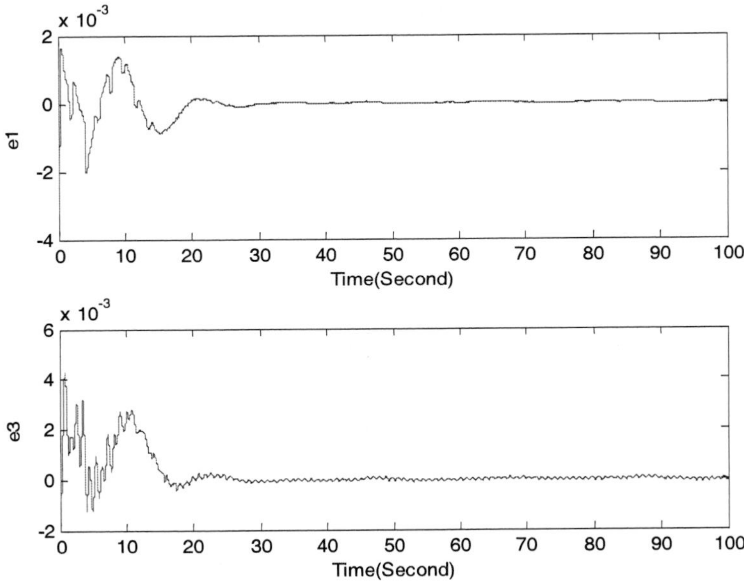

Figure 4.2. The tracking error using adaptive control.

At the beginning, using adaptive control, the estimate of angular velocity has a large amount overshoot and a much smaller rise time. The model uncertainties and external disturbances are difficult to compensate for in the adaptive controller because there is no disturbance term in the derivation; whereas, the disturbance term can be dealt with extensively in the adaptive sliding mode control. In the presence of model uncertainties and external disturbance, adaptive sliding mode control is, therefore, better than adaptive control.

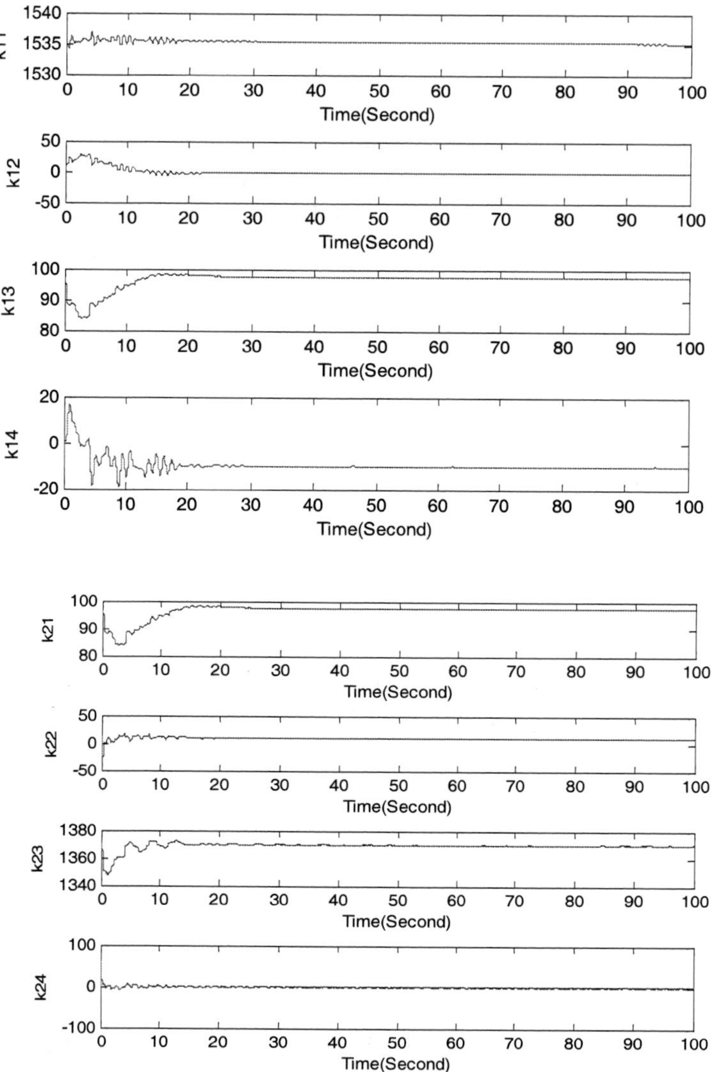

Figure 4.3. Adaptation of control parameters using adaptive control.

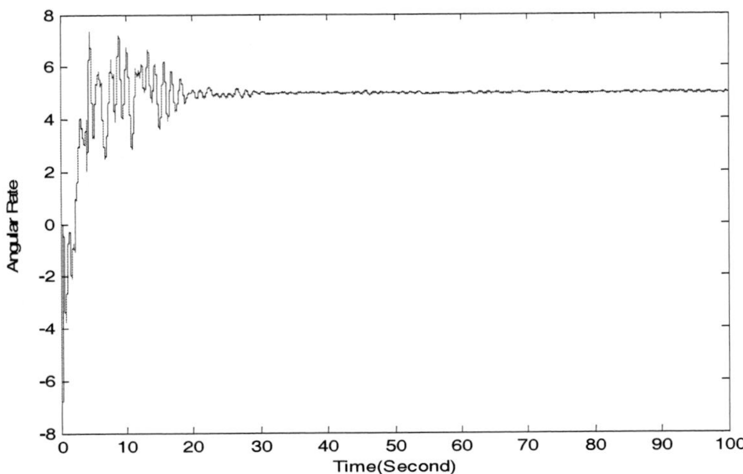

Figure 4.4. Adaptation of angular velocity using adaptive control.

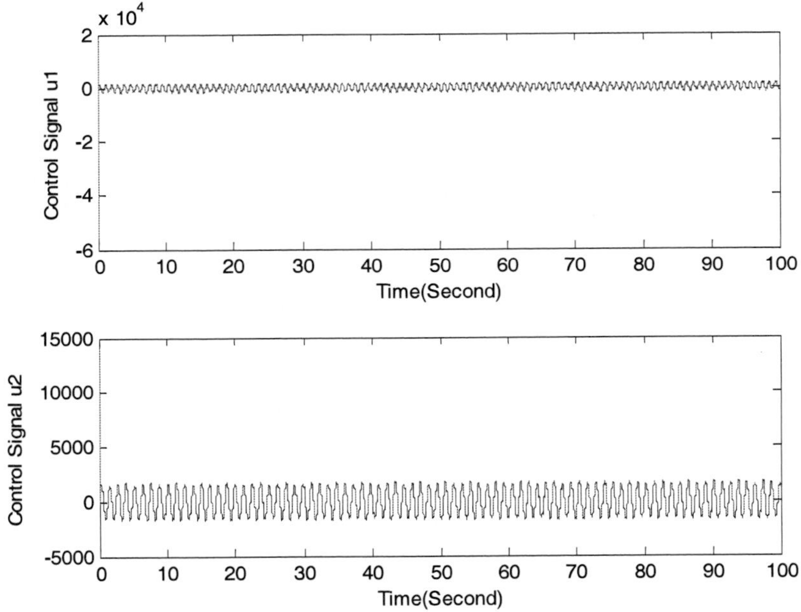

Figure 4.5. Adaptive control input.

CONCLUDING REMARKS

This chapter investigated the design of an adaptive control for the MEMS gyroscope. The dynamics model of the MEMS gyroscope was developed and nondimensionized. A novel adaptive approach was proposed and a stability condition was established. Simulation results demonstrated that the effectiveness of the proposed adaptive control techniques in identifying the gyroscope parameters and angular velocity.

Chapter 5

ROBUST ADAPTIVE CONTROL OF MEMS GYROSCOPES

This chapter presents a robust adaptive sliding mode controller for the MEMS gyroscope: an adaptive tracking controller with a proportional and integral sliding surface is proposed. The robust adaptive control algorithm can estimate the angular velocity vector and the linear damping and stiffness coefficients in real time. A robust adaptive controller, incorporating model uncertainties and external disturbances is derived in the Lyapunov sense and the stability of the closed-loop system is established. The numerical simulation is presented to verify the effectiveness of the proposed control scheme. It is exhibited that the proposed robust adaptive control scheme offers several advantages, such as a consistent estimation of gyroscope parameters (including angular velocity) and a large robustness to parameter variations and external disturbances. The motivation of this chapter is to propose a robust adaptive controller to estimate the angular velocity and all gyroscope parameters, including cross stiffness and damping coefficients in the presence of model uncertainties and external disturbances. A smooth sliding mode compensator is used to reduce control chattering. An adaptive law to update the parameters of adaptive sliding mode controller is also derived.

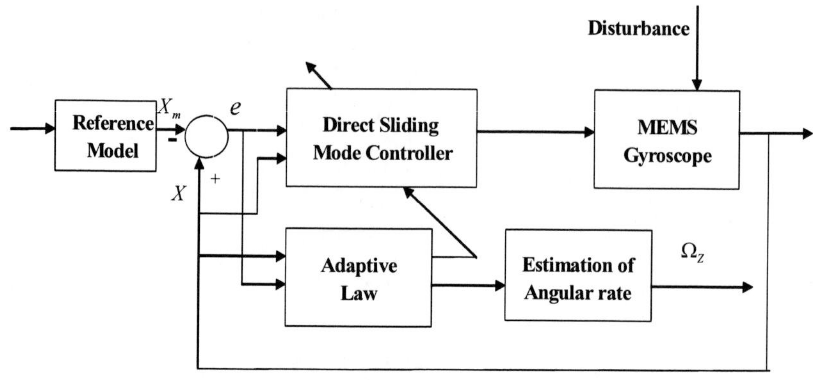

Figure 5.1. Block diagram of a direct adaptive sliding mode control.

5.1. Robust Adaptive Control Design

This section proposes an adaptive sliding mode control strategy for MEMS gyroscopes. The control target is to achieve real-time compensation for fabrication imperfections and a closed-loop identification of the angular velocity. The block diagram of an indirect adaptive sliding mode control for a MEMS gyroscope is shown in Figure 5.1:the tracking error between reference state and gyroscope state comes to the indirect adaptive sliding mode controller. The adaptive sliding mode controller is proposed to control the MEMS gyroscope. Angular velocity can be estimated by an adaptive estimator. Referring to (2.23), we consider the dynamics with parametric uncertainties and external disturbance as:

$$\ddot{q}+(D+2\Omega+\Delta D)\dot{q}+(K_b+\Delta K_b)q=u+d \tag{5.1}$$

where ΔD is the unknown parameter uncertainties of the matrix $D+2\Omega$, ΔK_b is the unknown parameter uncertainties of the matrix K_b, and d is an uncertain extraneous disturbance and/or unknown nonlinearity of the system .

Rewriting (5.1) as:

$$\ddot{q}+(D+2\Omega)\dot{q}+K_b q=u+f \tag{5.2}$$

where f represents the matched, lumped uncertainty and disturbance which is given by:

$$f = d - \Delta D\dot{q} - \Delta K_b q .$$

We make the following assumption:

The lumped uncertainty and disturbance f is bounded such as $\|f\| \le \alpha_1 \|q\| + \alpha_2 \|\dot{q}\| + \alpha_3$, where α_1, α_2 and α_3 are known positive constants .

Suppose that a reference trajectory is generated by an ideal oscillator and that the control objective is to make the trajectory of the gyroscopes to follow that of the reference model. The reference model is defined as:

$$\ddot{q}_m + K_m q_m = 0 \tag{5.3}$$

where $K_m = diag\{w_1{}^2 \quad w_2{}^2\}$.

The tracking error is defined as:

$$e = q - q_m . \tag{5.4}$$

The sliding surface is defined as:

$$s(t) = \dot{e} + \lambda e \tag{5.5}$$

where λ is a positive definite constant matrix to be selected, i.e. $\lambda = diag\{\lambda_1 \quad \lambda_2\}$.

The derivative of the sliding surface is:

$$\begin{aligned} \dot{s} &= \ddot{e} + \lambda \dot{e} \\ &= \ddot{q} - \ddot{q}_m + \lambda(\dot{q} - \dot{q}_m) \\ &= u + f - (D + 2\Omega)\dot{q} - K_b q + \lambda(\dot{q} - \dot{q}_m) + K_m q_m . \end{aligned} \tag{5.6}$$

Substituting $D=\begin{bmatrix} d_{xx} & d_{xy} \\ d_{xy} & d_{yy} \end{bmatrix}$, $\Omega=\begin{bmatrix} 0 & -\Omega_z \\ \Omega_z & 0 \end{bmatrix}$ and $K_b=\begin{bmatrix} w_x^{\ 2} & w_{xy} \\ w_{xy} & w_y^{\ 2} \end{bmatrix}$ into (5.6) yields:

$$\dot{s}=u+f-\begin{bmatrix} d_{xx} & d_{xy}-2\Omega_z \\ d_{xy}+2\Omega_z & d_{yy} \end{bmatrix}\begin{bmatrix} \dot{q}_1 \\ \dot{q}_2 \end{bmatrix}-\begin{bmatrix} w_x^{\ 2} & w_{xy} \\ w_{xy} & w_y^{\ 2} \end{bmatrix}\begin{bmatrix} q_1 \\ q_2 \end{bmatrix}+\lambda(\dot{q}-\dot{q}_m)+K_m q_m \,. \quad (5.7)$$

Rewriting (5.7) yields:

$$\dot{s}=u+f-\begin{bmatrix} \dot{q}_1 & \dot{q}_2 & 0 & -2\dot{q}_2 & q_1 & q_2 & 0 \\ 0 & \dot{q}_1 & \dot{q}_2 & 2\dot{q}_1 & 0 & q_1 & q_2 \end{bmatrix}\begin{bmatrix} d_{xx} \\ d_{xy} \\ d_{yy} \\ \Omega_z \\ w_x^{\ 2} \\ w_{xy} \\ w_y^{\ 2} \end{bmatrix}+\lambda(\dot{q}-\dot{q}_m)+K_m q_m \,. \quad (5.8)$$

Defining:

$$Y=\begin{bmatrix} \dot{q}_1 & \dot{q}_2 & 0 & -2\dot{q}_2 & q_1 & q_2 & 0 \\ 0 & \dot{q}_1 & \dot{q}_2 & 2\dot{q}_1 & 0 & q_1 & q_2 \end{bmatrix}, \quad (5.9)$$

$$\theta^*=\begin{bmatrix} d_{xx} & d_{xy} & d_{yy} & \Omega_z & w_x^{\ 2} & w_{xy} & w_y^{\ 2} \end{bmatrix}^T, \text{ and}$$

$$Q=\lambda(\dot{q}-\dot{q}_m)+K_m q_m \quad (5.10)$$

Then, (5.8) becomes:

$$\dot{s}=u+f-Y\theta^*+Q \quad (5.11)$$

where $Y(q,\dot{q},q_d,\dot{q}_d)$ is a 2×7 matrix of known functions and θ^* contains unknown system parameters. We assume both positions and velocities are measurable.

Setting $\dot{s}=0$ to solve equivalent control u_{eq} gives:

$$u_{eq}=Y\theta^*-Q-f\,. \tag{5.12}$$

The adaptive controller u is proposed as:

$$\begin{aligned} u &= Y\theta-Q+u_s \\ &= Y\theta-Q-\rho\,\mathrm{sgn}(s) \end{aligned} \tag{5.13}$$

where $u=\begin{bmatrix}u_1\\u_2\end{bmatrix}$, $s=\begin{bmatrix}s_1\\s_2\end{bmatrix}$, $u_s=\begin{bmatrix}u_{s1}\\u_{s2}\end{bmatrix}=-\rho\,\mathrm{sgn}(s)=-\begin{pmatrix}\rho_1 & 0\\0 & \rho_2\end{pmatrix}\begin{pmatrix}\mathrm{sgn}(s_1)\\\mathrm{sgn}(s_2)\end{pmatrix}$ is the sliding mode signal, and θ is the estimation of θ^*.

Substituting (5.13) into (5.11) yields:

$$\dot{s}=Y\tilde{\theta}+f-\rho\,\mathrm{sgn}(s) \tag{5.14}$$

where $\tilde{\theta}=\theta-\theta^*$.

Define a Lyapunov function to analyze the stability of (5.14) as:

$$V=\frac{1}{2}s^Ts+\frac{1}{2}\tilde{\theta}^T\tau^{-1}\tilde{\theta} \tag{5.15}$$

where $\tau=\tau^T$ is a positive definite matrix.

Differentiating V with respect to time yields:

$$\begin{aligned}\dot{V} &= s^T\dot{s} + \dot{\tilde{\theta}}^T\tau^{-1}\tilde{\theta} \\ &= s^T(Y\theta - Q - Y\theta^* + Q - \rho\,\mathrm{sgn}(s) + f) + \dot{\tilde{\theta}}^T\tau^{-1}\tilde{\theta} \\ &= s^T Y\tilde{\theta} - \rho(|s_1| + |s_2|) + s^T f + \dot{\tilde{\theta}}^T\tau^{-1}\tilde{\theta} \\ &= -\rho(|s_1| + |s_2|) + s^T f + (s^T Y\tilde{\theta} + \dot{\tilde{\theta}}^T\tau^{-1}\tilde{\theta}).\end{aligned} \quad (5.16)$$

To make $\dot{V} \le 0$, we choose an adaptive law:

$$\dot{\tilde{\theta}}(t) = \dot{\theta}(t) = -\tau Y^T s(t) \quad (5.17)$$

with $\theta(0)$ being arbitrary. This choice yields:

$$\begin{aligned}\dot{V} &= -\rho(|s_1| + |s_2|) + s^T f \le -\rho\|s\| + \|s\|\|f\| \\ &\le -\|s\|(\rho - \alpha_1\|q\| - \alpha_2\|\dot{q}\| - \alpha_3) \le 0.\end{aligned} \quad (5.18)$$

with the choice of $\rho \ge \alpha_1\|q\| + \alpha_2\|\dot{q}\| + \alpha_3 + \eta$, where η is a positive constant; and $\dot{V}$ becomes negative semi-definite, i.e. $\dot{V} \le -\eta\|s\|$. This implies that the trajectory reaches the sliding surface in finite time and remains on the sliding surface. Since $\dot{V}$ is negative definite, that implies that s and $\tilde{K}$ converge to zero. Since $\dot{V}$ is negative semi-definite, that ensures V, s and $\tilde{\theta}$ are all bounded. It can be concluded from (5.14) that $\dot{s}$ is also bounded.

Barbalat's lemma can be used to prove that $\lim_{t\to\infty} s(t) = 0$. The inequality $\dot{V} \le -\eta\|s\|$ implies that s is integrable as $\int_0^t \|s\| dt \le \frac{1}{\eta}[V(0) - V(t)]$. Since $V(0)$ is bounded and $V(t)$ is nonincreasing and bounded; it can be concluded that $\lim_{t\to\infty} \int_0^t \|s\| dt$ is bounded. Since $\lim_{t\to\infty} \int_0^t \|s\| dt$ is bounded and $\dot{s}$ is also bounded, according to Barbalat's lemma, $s(t)$ will asymptotically converge to zero, $\lim_{t\to\infty} s(t) = 0$.

Remark 1. The definition of Persistence of Excitation (PE)reads: a vector $v \in R^q \quad q \geq 1$ is said to have persistence of excitation if there exists positive constants α and T , such that for all $t > 0$, $\int_t^{t+T} v(\tau)v^T(\tau)d\tau \geq \alpha I$.

PE is a notion of a time signal that contains sufficient richness so that the $v(\tau)v^T(\tau)$ matrix is non-singular. It requires that $v(\tau)$ varies in such a way with time that the integral of the matrix is positive definite over any time interval $\left[t \quad t+T\right]$.

To make conclusions about $\widetilde{\theta} = 0$, other than the fact that they are bounded, we need to make the persistence of excitation argument. From the adaptive law $\dot{\widetilde{\theta}}(t) = \dot{\theta}(t) = -\tau Y^T s(t)$, according to [5]; if Y is persistently exciting a signal, then $\dot{\widetilde{\theta}}(t) = -\tau Y^T s(t)$ guarantees that $\widetilde{\theta} \to 0$, θ will converge to their true values. Because $s \to 0$ implies $e \to 0$, (5.4) determines that $q_1 = A_1 \sin(w_1 t)$, $\dot{q}_1 = w_1 A_1 \cos(w_1 t)$, $q_2 = A_2 \sin(w_2 t)$, and $\dot{q}_2 = w_2 A_2 \cos(w_2 t)$. It can illustrated that there exist some positive scalar constants, α and T ; such that for all $t > 0$, $\int_t^{t+T} Y^T Y d\tau \geq \alpha I$

where

$$Y^T Y = \begin{bmatrix} \dot{q}_1^2 & \dot{q}_1\dot{q}_2 & 0 & -2\dot{q}_1\dot{q}_2 & q_1\dot{q}_1 & \dot{q}_1 q_2 & 0 \\ \dot{q}_1\dot{q}_2 & \dot{q}_1^2 + \dot{q}_2^2 & \dot{q}_1\dot{q}_2 & -2\dot{q}_2^2 + 2\dot{q}_1^2 & q_1\dot{q}_2 & q_2\dot{q}_2 + q_1\dot{q}_1 & \dot{q}_1 q_2 \\ 0 & \dot{q}_2\dot{q}_1 & \dot{q}_2^2 & 2\dot{q}_1\dot{q}_2 & 0 & \dot{q}_2 q_1 & \dot{q}_2 q_2 \\ -2\dot{q}_1\dot{q}_2 & -2\dot{q}_2^2 + 2\dot{q}_1^2 & 2\dot{q}_1\dot{q}_2 & 4\dot{q}_2^2 + 4\dot{q}_1^2 & -2q_1\dot{q}_2 & -2q_2\dot{q}_2 + 2\dot{q}_1 q_2 & 2\dot{q}_1 q_2 \\ q_1\dot{q}_1 & q_1\dot{q}_2 & 0 & -2q_1\dot{q}_2 & q_1^2 & q_1 q_2 & 0 \\ q_2\dot{q}_1 & q_2\dot{q}_2 & 0 & -2q_2\dot{q}_2 & q_1 q_2 & q_2^2 & 0 \\ 0 & \dot{q}_1 q_2 & q_2\dot{q}_2 & -2q_2\dot{q}_1 & 0 & q_1 q_2 & q_2^2 \end{bmatrix}.$$

From (5.3) and (5.9) it can be shown that $Y^T Y$ has full rank if $w_1 \neq w_2$, i.e. the excitation frequencies on x and y axes, should be different. In other words, the excitation of proof mass should be persistent. Since $\widetilde{\theta} \to 0$, then the unknown angular velocity, as well as all other unknown parameters, can be estimated consistently. As a result, angular velocity Ω_z and all gyroscope parameters such as d_{xx} , d_{xy} , d_{yy} , w_x^2 , w_{xy} and w_y^2 converge to their true values.

In summary, if persistently exciting drive signals, $x_m = A_1 \sin(w_1 t)$ and $y_m = A_2 \sin(w_2 t)$ are used, then $\widetilde{\theta}(t)$, $s(t)$ and $e(t)$ all converge to zero, asymptotically. Consequently, the unknown angular velocity can be determined as: $\lim_{t\to\infty} \hat{\Omega}_z(t) = \Omega_z$; however, it is difficult to establish the convergence rate.

Remark 2. In the adaptive control system design, the persistent excitation condition is an important factor to estimate the angular velocity, Ω_z, correctly. The reference trajectory that the gyroscope must follow is generated in such a way that the resonance frequency of the x-axis is different from that of the y-axis,satisfing the persistent excitation condition.

5.2. Comparison to a Standard Adaptive Controller

A standard adaptive controller which has an addition term $K_f s$ is proposed as:

$$u = Y\theta - Q - K_f s \tag{5.19}$$

where θ is an estimate of θ^*, K_f is positive definite matrix.

Then (5.11) becomes:

$$\begin{aligned} \dot{s} &= u + f - Y\theta^* + Q \\ &= Y\widetilde{\theta} - K_f s + f. \end{aligned} \tag{5.20}$$

Define a Lyapunov function:

$$V = \frac{1}{2} s^T s + \frac{1}{2} \widetilde{\theta}^T \tau^{-1} \widetilde{\theta} \tag{5.21}$$

where $\tau = \tau^T > 0$, $\widetilde{\theta} = \theta - \theta^*$.

Differentiating V with respect to time yields:

$$\begin{aligned}\dot{V} &= s^T \dot{s} + \dot{\tilde{\theta}}^T \tau^{-1} \tilde{\theta} \\ &= s^T (Y\tilde{\theta} - K_f s + f) + \dot{\tilde{\theta}}^T \tau^{-1} \tilde{\theta} \\ &= -s^T K_f s + s^T f + (s^T Y \tilde{\theta} + \dot{\tilde{\theta}}^T \tau^{-1} \tilde{\theta}).\end{aligned} \tag{5.22}$$

To create $\dot{V} \leq 0$, we choose an adaptive law as:

$$\dot{\tilde{\theta}}(t) = \dot{\theta}(t) = -\tau Y^T s(t). \tag{5.23}$$

If $f = 0$; therefore, $\dot{V}$ becomes:

$$\dot{V} = -s^T K_f s \leq 0 \tag{5.24}$$

which implies that the stability of the closed-loop system can be guaranteed. If $f \neq 0$ than the stability of the closed-loop system cannot be guaranteed.

Remark 1. Such an adaptive controller would be inadequate to address the control system where appreciable non-parametric uncertainties exist, including: unmodelled dynamics, external disturbance and other imperfections in the estimates of gyroscopes parameters. Therefore, the proposed adaptive sliding mode controller incorporates the capability to maintain stable performance in the presence of model uncertainties and external disturbance.

Remark 2. In order to eliminate the control discontinuities, a smooth sliding mode control $\tanh(as)$ that can reduce the chattering problem is introduced. The parameter a determines the slope of the tanh(as) function at $s = 0$.

Therefore, the smooth sliding mode controller is proposed as :

$$u = Y\theta - Q - \rho \tanh(as). \tag{5.25}$$

5.3. Adaptive Sliding Mode Design under Asymmetric Coupling Term

If a gyroscope system does not have the same coupling damping and spring constant, the gyroscope system dynamics can be written as:

$$\ddot{q} + D\dot{q} + K_b q = u - 2\Omega\dot{q} \tag{5.26}$$

where $q = \begin{bmatrix} x \\ y \end{bmatrix}$ $u = \begin{bmatrix} u_x \\ u_y \end{bmatrix}$ $\Omega = \begin{bmatrix} 0 & -\Omega_z \\ \Omega_z & 0 \end{bmatrix}$, $D = \begin{bmatrix} d_{xx} & d_{xy} \\ d_{yx} & d_{yy} \end{bmatrix}$

$K_b = \begin{bmatrix} {w_x}^2 & w_{xy} \\ w_{yx} & {w_y}^2 \end{bmatrix}$.

The dynamics of sliding surface can be derived as:

$$\dot{s} = u + f - \begin{bmatrix} \dot{q}_1 & \dot{q}_2 & 0 & 0 & -2\dot{q}_2 & q_1 & q_2 & 0 & 0 \\ 0 & 0 & \dot{q}_1 & \dot{q}_2 & 2\dot{q}_1 & 0 & 0 & q_1 & q_2 \end{bmatrix} \begin{bmatrix} d_{xx} \\ d_{xy} \\ d_{yx} \\ d_{yy} \\ \Omega_z \\ w_x^2 \\ w_{xy} \\ w_{yx} \\ w_y^2 \end{bmatrix} + \lambda(\dot{q} - \dot{q}_m) + K_m q_m \tag{5.27}$$

Define:

$$Y = \begin{bmatrix} \dot{q}_1 & \dot{q}_2 & 0 & 0 & -2\dot{q}_2 & q_1 & q_2 & 0 & 0 \\ 0 & 0 & \dot{q}_1 & \dot{q}_2 & 2\dot{q}_1 & 0 & 0 & q_1 & q_2 \end{bmatrix} \tag{5.28}$$

$$\theta^* = \begin{bmatrix} d_{xx} & d_{xy} & d_{yx} & d_{yy} & \Omega_z & w_x^2 & w_{xy} & w_{yx} & w_y^2 \end{bmatrix}^T \tag{5.29}$$

$$Q = \lambda(\dot{q} - \dot{q}_m) + K_m q_m . \tag{5.30}$$

Similarly, as Lyapunov analysis before, the adaptive law is derived as:

$$\dot{\tilde{\theta}}(t) = \dot{\theta}(t) = -\tau Y^T s(t) . \tag{5.31}$$

Therefore, system parameters can be consistently estimated, including unsymmetrical coupling damping and spring parameters, such as: d_{xy}, d_{yx}, w_{xy} and w_{yx}.

Remark 1. The motion of a mode-unmatched gyroscope, in which the resonance frequency of the x-axis is different from that of the y-axis, has sufficient persistence of excitation to permit the identification of all major fabrication imperfections, as well as angular velocity. A MEMS gyroscope, suitable for the adaptive mode of operation, requires equal movements in the x and y axes. Thus, in the sense of conventional MEMS gyroscopes, there is no specific drive and sense axis. It should be noted that a conventional gyroscope structure is normally designed based on the assumption that the movement of the proof mass in the drive axis (x-axis) is relatively large, but the movement in the sense axis (y-axis) is very small. The proposed gyroscope design consists of a proof mass; four hairpin type spring suspensions; and for actuation and sensing located at both x and y axes, several pairs of parallel electrodes.

5.4. Simulation of a MEMS Gyroscope

We will evaluate the proposed adaptive sliding mode control with a sliding mode observer on the lumped MEMS gyroscope model. The control objective is to design an adaptive sliding mode controller so that the trajectory of $X(t)$ can track the state of reference model $X_m(t)$. In this simulation, we allowed $\pm$ 5% parameter variations for the spring and damping coefficients. Furthermore, we assumed $\pm$ 5% magnitude changes in the coupling terms, i.e. d_{xy} and ω_{xy}. The external disturbance is a random variable with zero mean and unit variance. Parameters of the MEMS gyroscope are as follows:

$m = 0.57e-8$ kg, $d_{xx} = 0.429e-6$ N s/m,

$d_{xy} = 0.0429e-6$ N s/m, $d_{yy} = 0.687e-6$ N s/m,

$k_{xx} = 80.98$ N/m, $k_{xy} = 5$ N/m, $k_{yy} = 71.62$ N/m,

$w_0 = 1kHz$, $q_0 = 10^{-6}m$.

The initial conditions $\theta(0) = 0.95\theta^*$, angular velocity $\Omega = 5.0$ rad/s. The desired motion trajectories are $x_m = \sin(w_1 t)$ and $y_m = 1.2\sin(w_2 t)$, where $w_1 = 4.17kHz$ and $w_2 = 5.11kHz$. The sliding gain is $\rho = diag\{200 \quad 200\}$, the adaptive gain is chosen as $\tau = diag\{200 \ 200 \ 200 \ 200 \ 200 \ 200 \ 200\}$. The sliding mode parameter is $\lambda = diag\{4 \quad 4\}$ and $a = 5$ in the smooth sliding mode controller tanh(as).

The tracking error and sliding surface are shown Figure 5.2 and Figure 5.3. It is illustrated that both tracking error and sliding surface converge to zero. Figure 5.4 and Figure 5.5 show that the estimation of spring and damping coefficients converge to their true values with a persistent sinusoidal reference signal. Figure 5.6 and Figure 5.7 compare the angular velocity estimation between smooth sliding mode controller and bang-bang type sliding mode controller sgn(s) under the sinusoidal reference signal. Both figures show that an estimation of angular velocity converges to its true values. With a smooth sliding mode controller, it has been observed that the estimated angular velocity has better convergence performance. Figure 5.8 depicts a smooth sliding mode control force of the adaptive sliding mode controller. It is shown from Figure 5.8 that an adaptive sliding mode system with a smooth sliding mode controller can significantly reduce chattering.

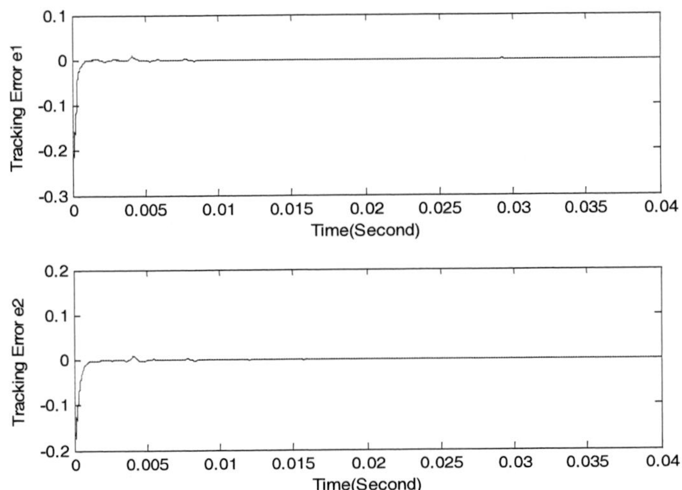

Figure 5.2. Convergence of the tracking error e(t).

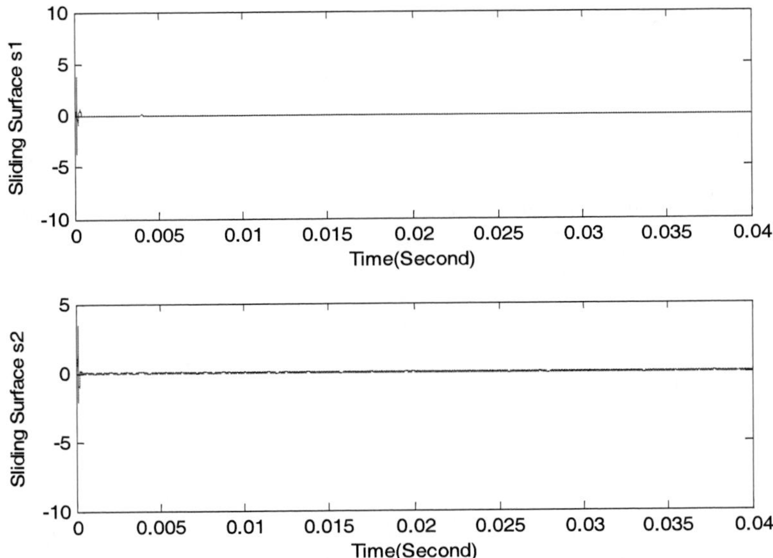

Figure 5.3. Convergence of the sliding surface s(t).

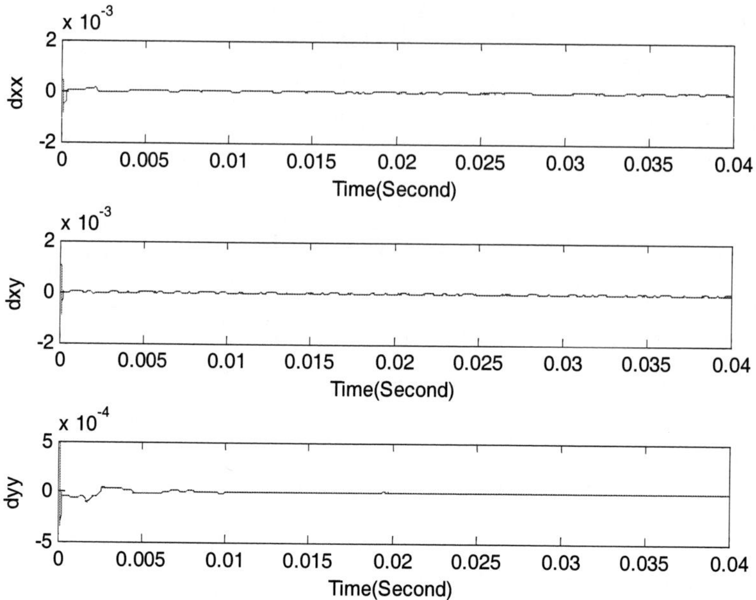

Figure 5.4. Adaptation of damping coefficients of gyroscope.

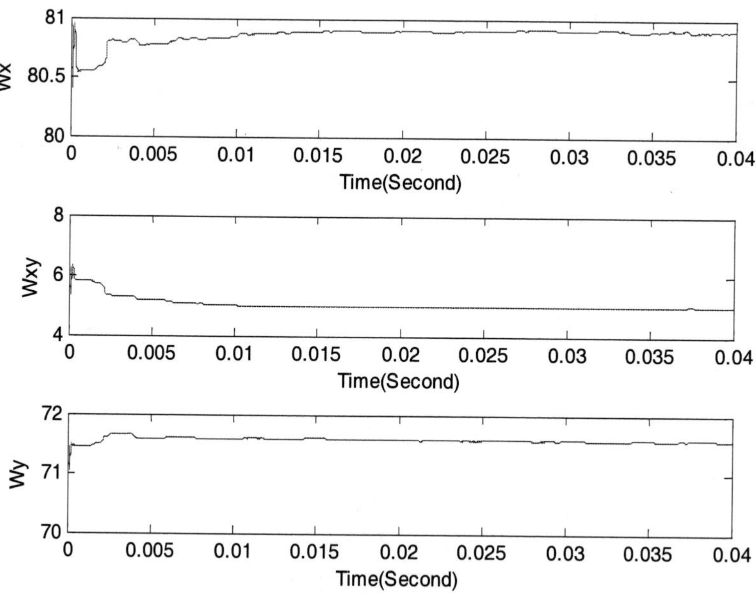

Figure 5.5. Adaptation of spring constants of gyroscope.

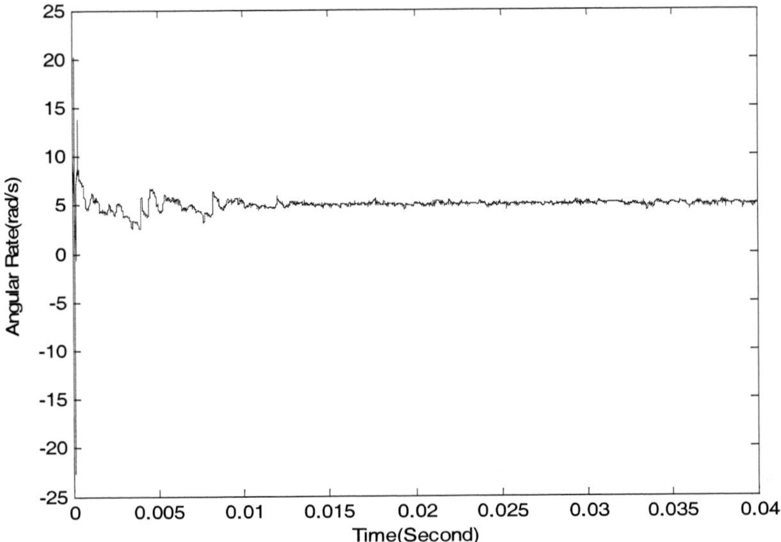

Figure 5.6. Convergence of the estimated angular velocity with smooth sliding mode controller.

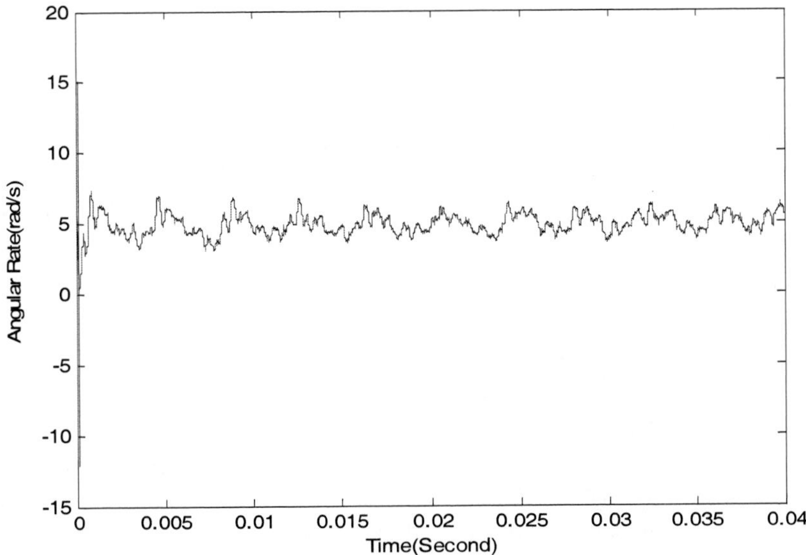

Figure 5.7. Convergence of the estimated angular velocity with bang-bang type sliding mode controller.

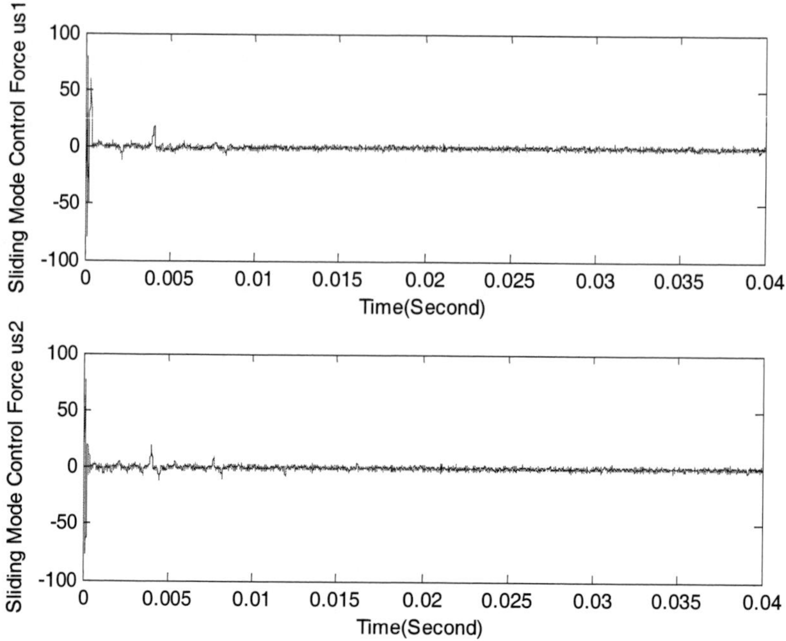

Figure 5.8. Smooth sliding mode control force of the adaptive sliding mode controller.

Concluding Remarks

This chapter investigated the design of robust adaptive control for the gyroscope. A new robust adaptive controller was formulated for MEMS gyroscopes, having two unmatched oscillatory modes having had sufficient persistence of excitation to permit the identification of all gyroscope parameters, including the damping and stiffness coefficients and angular velocity. The proposed robust adaptive controller incorporates the capability to maintain a stable performance in the presence of model uncertainties and external disturbance. Numerical simulations show that the proposed robust adaptive control has satisfactory performance and robustness in the presence of model uncertainty and external disturbance.

Chapter 6

ADAPTIVE SLIDING MODE CONTROL OF MEMS GYROSCOPES

This chapter presents an adaptive sliding mode tracking controller with a proportional and integral switching surface. A novel adaptive sliding mode controller, based on model reference adaptive state feedback control, is proposed to deal with the tracking problem for a class of dynamic systems. First, a proportional and integral sliding surface, instead of a conventional sliding surface is chosen; and, second, a class of adaptive sliding mode controller with integral sliding term is developed. It is shown that the stability of the closed-loop system can be guaranteed with the proposed adaptive sliding mode control strategy. Then, this chapter presents a new adaptive sliding mode controller, consisting of a proportional and integral sliding surface for the MEMS gyroscope. The adaptive sliding mode control algorithm can estimate the angular velocity vector and the linear damping and stiffness coefficients in real time. A proportional and integral sliding surface, instead of a conventional sliding surface is adopted. An adaptive sliding mode controller that incorporates both matched and unmatched uncertainties and disturbances is derived and the stability of the closed-loop system is established.

The design procedure of adaptive sliding mode controller consists of two steps: first, choosing a proportional, plus integral sliding surface; then deriving an adaptive sliding mode controller and establishing its stability. A novel adaptive control algorithm, associated with the integral sliding surface to update the parameters of the sliding mode controller, is also incorporated into the sliding mode control system. The feasibility of adaptive sliding mode control, with application to the vibratory gyroscope in the presence of the model uncertainties and external disturbances, is investigated. In this chapter, both matched and unmatched uncertainties are discussed within the same framework. The proposed adaptive sliding mode control, with proportional and integral sliding surface, can be applied to more a general model that is not solely limited to the gyroscope model. In this chapter, the adaptive control with a sliding mode compensator, used to identify the unknown controller parameters (with respect to the integral sliding surface) is developed, the latter having more application potential for a practical system.

The paramount contribution of this chapter will be its investigation into the feasibility of adaptive sliding mode control, in the presence of the model uncertainties and external disturbances, as it applies to both the time-varying linear model and the MEMS gyroscope. The chapter will also develop a more efficient controller: one that can not only identify all the gyroscope parameters and angular velocity, but with an improved robustness in the presence of the model uncertainties and external disturbances, as well. An adaptive law to estimate the parameters of the controller, with respect to the proposed integral sliding surface, is derived. The motivation is to propose a novel adaptive sliding mode controller with a proportional and integral switching surface. The proposed adaptive sliding mode controller is developed to compensate for the model uncertainties and input disturbances and to improve the tracking performance. A smooth sliding mode compensator is used to reduce control chattering. The numerical simulation of gyroscope is investigated to show the effectiveness of the proposed adaptive sliding mode control scheme with proportional and integral sliding mode surface.

6.1. Adaptive Sliding Mode Control Design

In this section, we will address the design of adaptive sliding mode control problem. Consider the system with multiple inputs with parametric uncertainties:

$$\dot{X}(t) = (A + \Delta A)X(t) + Bu(t) + f(t) \tag{6.1}$$

where $X(t) \in R^n$, $u(t) \in R^m$ and $A \in R^{n\times n}$, $B \in R^{n\times m}$ are unknown constant parameter matrices, ΔA is the unknown parameter uncertainties of the matrices A, $f(t)$ is an uncertain extraneous disturbance and nonlinearity of the system.

The reference model is given by:

$$\dot{X}_m(t) = A_m X_m(t) + B_m r(t) \tag{6.2}$$

where $X_m(t) \in R^n$, $r(t) \in R^m$, $A_m \in R^{n\times n}$, $B_m \in R^{n\times m}$ are known constant parameter matrices.

We make the following assumptions:

A1. All eigenvalues of A_m are in the open left-half complex plane, and $r(t)$ is bounded and piecewise continuous;

A2. There exists a constant matrix $K_1^* \in R^{n\times m}$ and a non-zero constant matrix $K_2^* \in R^{m\times m}$ such that the following equations are satisfied: $A + BK_1^{*T} = A_m$, $BK_2^* = B_m$;

A3. The sign of the λB is known, where $\lambda \in R^{m\times n}$ is a constant matrix such that λB is nonsingular.; $\|\lambda B\| < \mu$, where μ is known positive parameter, $\gamma_{\min}(\lambda B) > \delta$, where δ is positive parameter and $\gamma_{\min}(\lambda B)$ is the eigenvalue of matrix λB with minimum real part;

A4. There is a known matrix $Q \in R^{m\times m}$ such that $K_2^* Q$ is symmetric and positive definite:

$$M = K_2^* Q = (K_2^* Q)^T = Q^T K_2^{*T} > 0. \tag{6.3}$$

A5. ΔA and $f(t)$ have matched and unmatched terms. There exists unknown matrices of appropriate dimensions D, G such that $\Delta A(t) = BD(t) + \Delta\tilde{A}(t)$ and $f(t) = BG(t) + \tilde{f}(t)$, where $BD(t)$ is matched uncertainty and $\Delta\tilde{A}(t)$ is unmatched uncertainty, $BG(t)$ is matched disturbance and $\tilde{f}(t)$ is unmatched disturbance.

From this assumption, equation (6.1) can be rewritten as:

$$\dot{X}(t) = AX(t) + Bu(t) + \Delta AX(t) + f(t)$$
$$= AX(t) + Bu(t) + BDX(t) + \Delta\tilde{A}x(t) + BG + \tilde{f}(t) \quad (6.4)$$
$$= AX(t) + Bu(t) + Bf_m + f_u$$

where $f_m(t, X)$ represents the system's matched, lumped uncertainty and disturbance, lying in the range space of *B*, is given by:

$$f_m(t, X) = DX(t) + G \quad (6.5)$$

and where $f_u(t, X)$ represents the system's lumped, unmatched uncertainty and disturbance, is given by:

$$f_u(t, X) = \Delta\tilde{A}X(t) + \tilde{f}(t). \quad (6.6)$$

A6. The matched and unmatched lumped uncertainty and disturbance f_m and f_u are bounded by known positive parameters α_{m1}, α_{m2} and α_{u1}, α_{u2} such as:

$$\|f_m(t, X)\| \le \alpha_{m1}\|X\| + \alpha_{m2} \text{ and } \|f_u(t, X)\| \le \alpha_{u1}\|X\| + \alpha_{u2}.$$

The tracking error is defined as $e(t) = X(t) - X_m(t)$, the derivative of tracking error is:

$$\begin{aligned} \dot{e} &= AX + Bu + Bf_m + f_u - (A_m X_m + B_m r) \\ &= A_m e + (A - A_m)X + Bu + Bf_m + f_u - B_m r. \end{aligned} \quad (6.7)$$

The integral sliding surface is defined as:

$$s(t) = \lambda e - \int_0^t \lambda A_m e d\tau \quad (6.8)$$

The derivative of the sliding surface is:

$$\dot{s} = \lambda A_m e + \lambda(A - A_m)X + \lambda Bu + \lambda Bf_m + \lambda f_u - \lambda B_m r - \lambda A_m e \\ = \lambda(A - A_m)X + \lambda Bu + \lambda Bf_m + \lambda f_u - \lambda B_m r. \tag{6.9}$$

Setting $\dot{s} = 0$ to solve equivalent control u_{eq} gives:

$$u_{eq} = -(\lambda B)^{-1}\lambda[(A - A_m)X - B_m r] - f_m - (\lambda B)^{-1}\lambda f_u \\ = -(\lambda B)^{-1}\lambda(A - A_m)X + (\lambda B)^{-1}\lambda B_m r - f_m - (\lambda B)^{-1}\lambda f_u \\ = K_1^{*T} X(t) + K_2^* r(t) - f_m - (\lambda B)^{-1}\lambda f_u. \tag{6.10}$$

where $K_1^* = (\lambda B)^{-1}\lambda(A_m - A)$, $K_2^* = (\lambda B)^{-1}\lambda B_m$.

Remark 1. When the system is in the sliding mode, it is known that $s = 0$ and $\dot{s} = 0$, and there exists an equivalent controller such that $\dot{s}(t) = 0$. So the closed-loop error dynamic equation, after the system enters the sliding surface is $\dot{e}(t) = A_m e(t) + (I - B(\lambda B)^{-1}) f_u$.

If the unmatched uncertainty and disturbance $f_u = 0$, the error dynamics in the sliding surface is $\dot{e}(t) = A_m e(t)$. therefore, the sliding dynamics could be influenced by the unmatched uncertainties.

The adaptive control signal u is proposed as:

$$u(t) = K_1^T(t)X(t) + K_2(t)r(t) - \rho\frac{s}{\|s\|} \tag{6.11}$$

where $K_1(t)$ and $K_2(t)$ are the estimates of K_1^* and K_2^* respectively, ρ is constant, $\|\cdot\|$ is the Euclidean norm, $\rho\frac{s}{\|s\|}$ is the unit sliding mode signal.

Define the parameter errors as:

$$\tilde{K}_1(t) = K_1(t) - K_1^* \tag{6.12}$$

$$\tilde{K}_2(t) = K_2(t) - K_2^* . \tag{6.13}$$

Substituting (6.11), (6.12), (6.13) into (6.1), we get:

$$\begin{aligned}\dot{X}(t) &= AX(t) + B[K_1^T(t)X(t) + K_2(t)r(t)] + Bf_m + f_u - B\rho\frac{s}{\|s\|} \\ &= A_m X(t) + B_m r(t) + B_m\left[K_2^{*-1}\tilde{K}_1^T(t)X(t) + K_2^{*-1}\tilde{K}_2(t)r(t)\right] + Bf_m + f_u - B\rho\frac{s}{\|s\|}.\end{aligned} \tag{6.14}$$

Substituting (6.11), (6.14) into (6.6), yields a tracking error equation:

$$\dot{e}(t) = A_m e(t) + B_m\left[K_2^{*-1}\tilde{K}_1^T(t)X(t) + K_2^{*-1}\tilde{K}_2(t)r(t)\right] + Bf_m + f_u - B\rho\frac{s}{\|s\|} \tag{6.15}$$

and the derivative of $s(t)$ is:

$$\dot{s}(t) = \lambda B_m\left[K_2^{*-1}\tilde{K}_1^T(t)X(t) + K_2^{*-1}\tilde{K}_2(t)r(t)\right] + \lambda Bf_m + \lambda f_u - \lambda B\rho\frac{s}{\|s\|} . \tag{6.16}$$

Since the adaptive laws for $K_1(t)$ and $K_2(t)$ are to be chosen as the dynamic, the state vector of the closed-loop system is defined as:

$$e_c(t) = \left(s^T(t), \tilde{K}_1^T(t), \tilde{K}_2(t)\right)^T . \tag{6.17}$$

Define a Lyapunov function:

$$V(e_c) = \frac{1}{2}s^T s + \frac{1}{2}tr[\tilde{K}_1^T M^{-1}\tilde{K}_1] + \frac{1}{2}tr[\tilde{K}_2^{\,T} M^{-1}\tilde{K}_2] \tag{6.18}$$

where $M = M^T > 0$ satisfies the assumption A4, $tr[M]$ denotes the trace of a square matrix M.

Differentiating V, with respect to time, yields:

$$\begin{aligned}
\dot{V} &= s^T \dot{s} + tr[\widetilde{K}_1 M^{-1} \dot{\widetilde{K}}_1^T] + tr[\widetilde{K}_2^T M^{-1} \dot{\widetilde{K}}_2] \\
&= s^T \lambda B_m \left[K_2^{*-1} \widetilde{K}_1^T(t) X(t) + K_2^{*-1} \widetilde{K}_2(t) r(t) \right] \\
&- s^T \lambda B \rho \frac{s}{\|s\|} + s^T \lambda B f_m + s^T \lambda f_u + tr[\widetilde{K}_1 M^{-1} \dot{\widetilde{K}}_1^T] + tr[\widetilde{K}_2^T M^{-1} \dot{\widetilde{K}}_2] \\
&= -s^T \lambda B \rho \frac{s}{\|s\|} + s^T \lambda B f_m + s^T \lambda f_u + \left[tr[\widetilde{K}_1 M^{-1} Q^T B_m^T \lambda^T s^T X^T] + tr[\widetilde{K}_1 M^{-1} \dot{\widetilde{K}}_1^T] \right] \\
&+ \left[tr[\widetilde{K}_2^T M^{-1} Q^T B_m^T \lambda^T s^T r^T] + tr[\widetilde{K}_2^T M^{-1} \dot{\widetilde{K}}_2] \right]
\end{aligned} \tag{6.19}$$

where we use the definition $M = K_2^* Q = M^T > 0$ and the properties $tr[N_1 N_2] = tr[N_2 N_1]$, $tr[N_3] = tr[N_3^T]$ for any matrices N_1, N_2 and N_3, where $tr[N]$ denotes the trace of a square matrix N.

To make $\dot{V} \leq 0$, we choose the adaptive laws as:

$$\dot{\widetilde{K}}_1^T(t) = \dot{K}_1^T(t) = -Q^T B_m^T \lambda^T s^T X^T \tag{6.20}$$

$$\dot{\widetilde{K}}_2(t) = \dot{K}_2(t) = -Q^T B_m^T \lambda^T s^T r^T \tag{6.21}$$

with Q satisfying assumption A4 and $K_1(0)$ and $K_2(0)$ being arbitrary. With this adaptive law choice, $\gamma_{\min}(\lambda B) > \delta$ and $\rho > \dfrac{\mu(\alpha_{m1}\|x\| + \alpha_{m2}) + \|\lambda\|(\alpha_{u1}\|x\| + \alpha_{u1})}{\delta}$, we have:

$$\begin{aligned}
\dot{V} &= -s^T \lambda B \rho \frac{s}{\|s\|} + s^T \lambda B f_m + s^T \lambda f_u \\
&\leq -\rho \gamma_{\min}(\lambda B)\|s\| + \|s\|\|\lambda B\|\|f_m\| + \|s\|\|\lambda\|\|f_u\| \\
&\leq -\rho \gamma_{\min}(\lambda B)\|s\| + \mu\|s\|\|f_m\| + \|s\|\|\lambda\|\|f_u\| \\
&< -\gamma_{\min}(\lambda B)\|s\| \left[\rho - \frac{\mu(\alpha_{m1}\|x\| + \alpha_{m2}) + \|\lambda\|(\alpha_{u1}\|x\| + \alpha_{u1})}{\delta} \right] \\
&< 0
\end{aligned} \tag{6.22}$$

where we make use the property of $\gamma_{\min}(\lambda B)\|s\|^2 \leq s^T \lambda B s \leq \gamma_{\max}(\lambda B)\|s\|^2$.

$\dot{V} < 0$ implies that the equilibrium state $e_c = 0$ of the closed-loop system is uniformly stable, as long as $s \neq 0$ and its solution $e_c(t) \to 0$. This implies $\dot{V}$ is negative semi definite; s, $\widetilde{K}_1$ and $\widetilde{K}_2$ are bounded; from [16], we can see $\dot{s}$ is also bounded; $\ddot{V}$ is bounded, implying that $\dot{V}$ is uniformly continuous, and according to barbalat lemma, $\dot{V}$ converges to zero as time goes on. Thus, we conclude that $\lim_{t\to\infty} s(t) = 0$, $s(t)$ asymptotically converges to zero.

From the adaptive laws (6.20) and (6.21), according to the persistence excitation theory, if X and r are persistent excitation signals, then $\dot{\widetilde{K}}_1^T(t) = -Q^T B_m^T \lambda^T s^T X^T$ and $\dot{\widetilde{K}}_2(t) = -Q^T B_m^T \lambda^T s^T r^T$ guarantee that $\widetilde{K}_1 \to 0$, $\widetilde{K}_2 \to 0$; meaning that K_1 and K_2 will converge to their true values. It can be concluded that the controller parameters converge to their true values if the condition of persistent excitation can be satisfied.

Remark 2. In order to eliminate the chattering, the discontinuous control component in (6.11) can be replaced by a smooth sliding mode component to yield:

$$u(t) = K_1^T(t)X(t) + K_2 r(t) - \rho \frac{s}{\|s\| + \varepsilon} \tag{6.23}$$

where $\varepsilon > 0$ is small constant. This creates a small boundary layer about the switching surface, in which the system trajectory will remain. Therefore, the chattering problem can be reduced significantly.

6.2. Application to the MEMS Gyroscope

We consider a MEMS gyroscope model and evaluate the proposed adaptive sliding mode control with proportional and integral sliding surface.

Remark 3.1 Because Bm=0 for the reference model of the MEMS gyroscope, we, therefore, modify the assumption A2 in section 6.1 as: there exists a constant matrix K^*, such that the following matching condition $A + BK^{*T} = A_m$ can always be satisfied. We also ignore the assumption A4 in section 6.1. Now, there is little difference between the proposed adaptive sliding mode control theory for the general case and the special numerical example of the triaxial MEMS gyroscope.

This section proposes a new direct adaptive sliding mode control strategy with a proportional and integral sliding surface for MEMS gyroscopes. A detailed study of the proportional-integral sliding mode control algorithm is presented and solved in the presence of both matched and mismatched model uncertainties and disturbances. The control target for MEMS gyroscope is to design an adaptive sliding mode controller so that the trajectory of the driving axes can track the state of a reference model. The purpose of the sliding mode control is to establish conditions that the unknown angular velocity Ω_z can be consistently estimated.

For compactness, the gyroscope model is written in state space form as:

$$\dot{X} = AX + Bu \tag{6.24}$$

where

$$A = \begin{bmatrix} 0 & 1 & 0 & 0 \\ -w_x^{\ 2} & -d_{xx} & -w_{xy} & -(d_{xy} - 2\Omega_z) \\ 0 & 0 & 0 & 1 \\ -w_{xy} & -(d_{xy} + 2\Omega_z) & -w_y^{\ 2} & -d_{yy} \end{bmatrix}$$

$$B = \begin{bmatrix} 0 & 1 & 0 & 0 \\ 0 & 0 & 0 & 1 \end{bmatrix}^T, \quad u = \begin{bmatrix} u_x \\ u_y \end{bmatrix}, \quad X = \begin{bmatrix} x \\ \dot{x} \\ y \\ \dot{y} \end{bmatrix}.$$

The reference models $x_m = A_1 \sin(w_1 t)$, $y_m = A_2 \sin(w_2 t)$ are defined by:

$$\ddot{q}_m + K_m q_m = 0 \tag{6.25}$$

where $K_m = diag\{w_1^2 \quad w_2^2\}$. Similar to (6.24), the reference model can be written as:

$$\dot{X}_m = \begin{bmatrix} 0 & 1 & 0 & 0 \\ -w_1^2 & 0 & 0 & 0 \\ 0 & 0 & 0 & 1 \\ 0 & 0 & -w_2^2 & 0 \end{bmatrix} X_m \equiv A_m X_m \qquad (6.26)$$

where A_m is a known constant matrix.

Consider the system in (6.24) with parametric uncertainties ΔA and external disturbance f as:

$$\dot{X}(t) = (A + \Delta A)X(t) + Bu + f(t) \qquad (6.27)$$

where $x(t) \in R^n$, $u(t) \in R^m$ and $A \in R^{n \times n}$ is the unknown constant matrix, ΔA is the unknown uncertainties of the matrix A, $f(t)$ is an uncertain extraneous disturbance .

We make the following assumptions:

A1. ΔA and $f(t)$ have matched and unmatched terms. There exist unknown matrices of appropriate dimensions D , G , such that $\Delta A(t) = BD(t) + \Delta\tilde{A}(t)$ and $f(t) = BG(t) + \tilde{f}(t)$, where $BD(t)$ is a matched uncertainty and $\Delta\tilde{A}(t)$ is an unmatched uncertainty. $BG(t)$ is a matched disturbance and $\tilde{f}(t)$ is an unmatched disturbance. Therefore, the dynamics (6.27) can be rewritten as:

$$\dot{X}(t) = AX(t) + Bu(t) + Bf_m + f_u \qquad (6.28)$$

where $Bf_m(t, X)$ represents the system lumped matched uncertainty and disturbance, given by:

$$f_m(t, X) = DX(t) + G\,. \qquad (6.29)$$

The term $f_u(t,X)$ represents the lumped unmatched uncertainty and disturbance which takes the form:

$$f_u(t,X) = \Delta\widetilde{A}X(t) + \widetilde{f}(t). \tag{6.30}$$

A2. The matched and unmatched lumped uncertainty and external disturbance f_m and f_u are bounded such as $\|f_m(t,X)\| \le \alpha_m$ and $\|f_u(t,X)\| \le \alpha_u$, where α_m and α_u, are known positive constants , where $\|\cdot\|$ is the Euclidean norm.

A3. There exists a constant matrix K^*, such that the matching condition, $A + BK^{*T} = A_m$, can always be satisfied.

The tracking error and its derivative are:

$$e(t) = X(t) - X_m(t) \tag{6.31}$$

$$\dot{e} = A_m e + (A - A_m)X + Bu + Bf_m + f_u . \tag{6.32}$$

The proportional-integral sliding surface $s = 0$ s defined as:

$$s(t) = \lambda e - \int_0^t \lambda(A_m + BK_e)ed\tau \tag{6.33}$$

where λ is a constant matrix such that λB is nonsingular. The constant K_e satisfies the condition that $(A_m + BK_e)$ is Hurwitz .

The derivative of the sliding surface is:

$$\dot{s} = \lambda(A - A_m)X + \lambda Bu + \lambda Bf_m + \lambda f_u - \lambda BK_e e . \tag{6.34}$$

Setting $\dot{s} = 0$ to solve equivalent control u_{eq} gives:

$$u_{eq} = -(\lambda B)^{-1}\lambda(A - A_m)X + K_e e - f_m - (\lambda B)^{-1}\lambda f_u \\ = K^{*T}X(t) + K_e e - f_m - (\lambda B)^{-1}\lambda f_u . \tag{6.35}$$

An adaptive version of the control algorithm is proposed as:

$$u(t) = K^T(t)X(t) + K_e e(t) - \rho(\lambda B)^{-1}\frac{s}{\|s\|} \tag{6.36}$$

where $K(t)$ is an estimate of K^*. The last component of the control signal is designed to address the matched and unmatched disturbances.

This component is given as: $u_s = \begin{bmatrix} u_{s1} \\ u_{s2} \end{bmatrix} = -\rho(\lambda B)^{-1}\frac{s}{\|s\|}$, where ρ is a constant.

We define the estimation error as:

$$\widetilde{K}(t) = K(t) - K^* . \tag{6.37}$$

Substituting (6.37) and (6.36) into (6.28) yields:

$$\dot{X}(t) = A_m X(t) + B\widetilde{K}^T(t)X(t) + BK_e e + Bf_m + f_u - B\rho(\lambda B)^{-1}\frac{s}{\|s\|}. \tag{6.38}$$

The tracking error equation now becomes:

$$\dot{e}(t) = (A_m + BK_e)e + B\widetilde{K}^T(t)X(t) + Bf_m + f_u - B\rho(\lambda B)^{-1}\frac{s}{\|s\|}. \tag{6.39}$$

The dynamics of sliding surface $s(t)$ is:

$$\dot{s}(t) = \lambda B\widetilde{K}^T(t)X(t) + \lambda Bf_m + \lambda f_u - \rho\frac{s}{\|s\|}. \tag{6.40}$$

Define a Lyapunov function as:

$$V = \frac{1}{2} s^T s + \frac{1}{2} tr[\tilde{K} M^{-1} \tilde{K}^T] \tag{6.41}$$

where $M = diag\{m_1 \quad m_2\}$ is positive definite matrix.

Differentiating V with respect to time yields:

$$\begin{aligned}
\dot{V} &= s^T \dot{s} + tr[\tilde{K} M^{-1} \dot{\tilde{K}}^T] \\
&= s^T \left[\lambda B \tilde{K}^T(t) X(t) + \lambda B f_m + \lambda f_u - \rho \frac{s}{\|s\|} \right] + tr[\tilde{K} M^{-1} \dot{\tilde{K}}^T] \\
&= -s^T \rho \frac{s}{\|s\|} + s^T \lambda B f_m + s^T \lambda f_u + s^T \lambda B \tilde{K}^T(t) X(t) + tr[\tilde{K} M^{-1} \dot{\tilde{K}}^T].
\end{aligned} \tag{6.42}$$

To make $\dot{V} \leq 0$, we choose the adaptive law as:

$$\dot{\tilde{K}}^T(t) = \dot{K}^T(t) = -M B^T \lambda^T s X^T(t) \tag{6.43}$$

with $K(0)$ being arbitrary. This adaptive law yields:

$$\begin{aligned}
\dot{V} &= -\rho\|s\| + s^T \lambda B f_m + s^T \lambda f_u \\
&\leq -\rho\|s\| + \|s\|\|\lambda B\|\|f_m\| + \|s\|\|\lambda\|\|f_u\| \\
&\leq -\rho\|s\| + \|s\|\|\lambda B\|\alpha_m + \|s\|\|\lambda\|\alpha_u \\
&= -\|s\|(\rho - \|\lambda B\|\alpha_m - \|\lambda\|(\alpha_{u1}).
\end{aligned} \tag{6.44}$$

with $\rho \geq \|\lambda B\|\alpha_m + \|\lambda\|\alpha_u + \eta$, where η is a positive constant, $\dot{V}$ becomes negative semi-definite, i.e., $\dot{V} \leq -\eta\|s\|$. This implies that the trajectory reaches the sliding surface in finite time and remains on the sliding surface. $\dot{V}$ is negative definite, implying that s and $\tilde{K}$ converge to zero. $\dot{V}$ is negative semi-definite, ensuring V, s and $\tilde{K}$ are all bounded. It can be concluded from (6.40) that $\dot{s}$ is also bounded. The inequality $\dot{V} \leq -\eta\|s\|$ implies that s is integrable as $\int_0^t \|s\| dt \leq \frac{1}{\eta}[V(0) - V(t)]$. Since $V(0)$ is bounded and $V(t)$ is nonincreasing and bounded, it can be concluded that $\lim_{t\to\infty}\int_0^t \|s\| dt$ is bounded. Since $\lim_{t\to\infty}\int_0^t \|s\| dt$ is bounded and $\dot{s}$ is also bounded, according to Barbalat lemma, $s(t)$ will asymptotically converge to zero, $\lim_{t\to\infty} s(t) = 0$. Then, $e(t)$ will also asymptotically converge to zero. From the adaptive laws $\dot{\tilde{K}}^T(t) = \dot{K}^T(t) = -MB^T\lambda^T sX^T$; according to the persistence excitation theory, if X is persistent excitation signal, then $\dot{\tilde{K}}^T(t) = -MB^T\lambda^T sX^T$ guarantees that $\tilde{K} \to 0$, K will converges to its true value. It can be shown that if $w_1 \neq w_2$, there always exists some positive scalar constants α and T , such that for all $t > 0$, $\int_t^{t+T} XX^T d\tau \geq \alpha I$,

where $XX^T = \begin{bmatrix} x_1^2 & x_1\dot{x}_1 & x_1x_2 & x_1\dot{x}_2 \\ \dot{x}_1x_1 & \dot{x}_1^2 & \dot{x}_1x_2 & \dot{x}_1\dot{x}_2 \\ x_2x_1 & x_2\dot{x}_1 & x_2^2 & x_2\dot{x}_2 \\ \dot{x}_2x_1 & \dot{x}_2\dot{x}_1 & \dot{x}_2x_2 & \dot{x}_2^2 \end{bmatrix}$.

It can be shown that XX^T has full rank if $w_1 \neq w_2$, i.e. the excitation frequencies on x and y axes should be different. In other words, excitation of the proof mass should be persistently exciting. Since $\tilde{K} \to 0$, then the unknown angular velocity, as well as all other unknown parameters, can be determined from $A + BK^T = A_m$.

In summary, if persistently exciting driven signals, $x_m = A_1 \sin(w_1 t)$ and $y_m = A_2 \sin(w_2 t)$ are used, then $\tilde{K}$, $s(t)$ and $e(t)$ all converge to zero asymptotically. Consequently the unknown angular velocity can be determined as $\lim_{t\to\infty}\hat{\Omega}_z(t)=\Omega_z$. It is difficult, however, to establish the convergence rate.

Remark 1. If (A_m, B) is a controllable pair, the closed-loop matrix $A_m + BK_e$ can be assigned to have arbitrary set of eigenvalues to control the tracking error settling rates. Therefore, K_e is chosen such that $(A_m + BK_e)$ is Hurwitz .

Remark 2. In order to eliminate the chattering, the discontinuous control component can be replaced by a smooth sliding mode component to yield:

$$u(t)=K^T(t)X(t)+K_e e-(\lambda B)^{-1}\rho\frac{s}{\|s\|+\varepsilon} \tag{6.45}$$

where $\varepsilon > 0$ is a small constant. This modification creates a small boundary layer around the switching surface in which the system trajectory remains. Therefore, the chattering problem can be significantly reduced.

6.3. Simulation Example

We evaluated the proposed adaptive sliding mode control on a lumped MEMS gyroscope model. The control objective is to design an adaptive sliding mode controller so that a consistent estimate of Ω_z can be obtained.

In the simulation, with respect to their nominal values, $\pm$ 5% parameter variations were allowed for the spring and damping coefficients, $\pm$ 2% magnitude changes in the coupling terms d_{xy} and ω_{xy} were assumed. The $\pm$ 5% time varying parameter variation of Ω_z was also assumed. All the parameter variations above are included in the uncertainties of the system parameters ΔA . In the simulation, it is assumed that only ΔA has an unmatched term and the external disturbance $f(t)$ has both matched and unmatched terms. Both matched and unmatched disturbance are assumed to be a random variable, with zero mean and unity variance. Parameters of the MEMS gyroscope are as follows:

$m = 0.57e-8$ kg, $d_{xx} = 0.429e-6$ N s/m,

$d_{xy} = 0.0429e-6$ N s/m, $d_{yy} = 0.687e-6$ N s/m,

$k_{xx} = 80.98$ N/m, $k_{xy} = 5$ N/m, $k_{yy} = 71.62$ N/m,

$w_0 = 1kHz$, $q_0 = 10^{-6} m$.

The unknown angular velocity Ω_z is assumed to be a step function of the amplitude 5.0 rad/s, and the initial condition on the K matrix is $K(0) = 0.95K^*$. The desired motion trajectories are $x_m = \sin(w_1 t)$ and $y_m = 1.2\sin(w_2 t)$, where $w_1 = 4.17kHz$ and $w_2 = 5.11kHz$. K_e and sliding matrix λ in (6.36) are chosen as: $K_e = \begin{bmatrix} -10000 & -10000 & 1000 & 20000 \\ -1000 & -1000 & -1000 & -1000 \end{bmatrix}$ and $\lambda = \begin{bmatrix} 0 & 10 & 0 & 0 \\ 0 & 0 & 0 & 10 \end{bmatrix}$ respectively. The sliding mode gain of (6.36) is chosen as $\rho = diag\{10000 \quad 10000\}$ to satisfy the stability condition. The adaptive gain of (6.43) is $M = diag\{5 \quad 5\}$. The smooth sliding mode component of (6.45) has $\varepsilon = 0.1$.

The plots of tracking error and sliding surface are shown in Figures 6.1-6.2. Figure 6.3 demonstrates that the estimation of controller parameters converge to their true values and with persistently exciting driving signals on both axes.

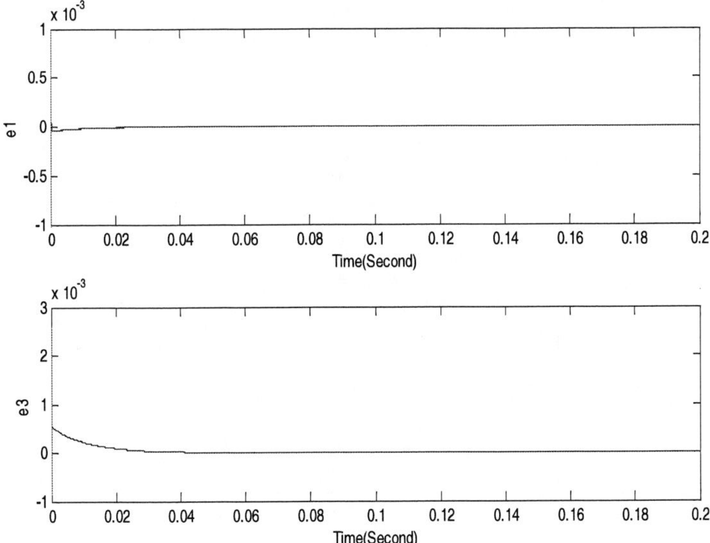

Figure 6.1. Plot of the tracking error e(t).

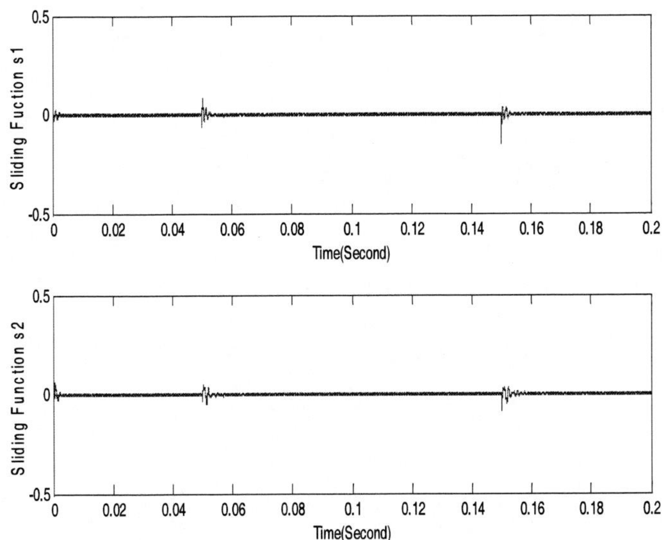

Figure 6.2. Plot of the sliding surface s(t).

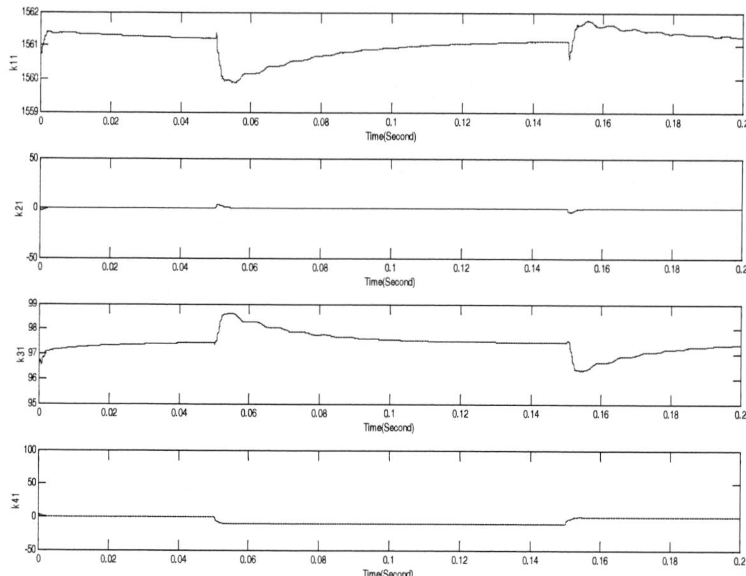

Figure 6.3. (Continued).

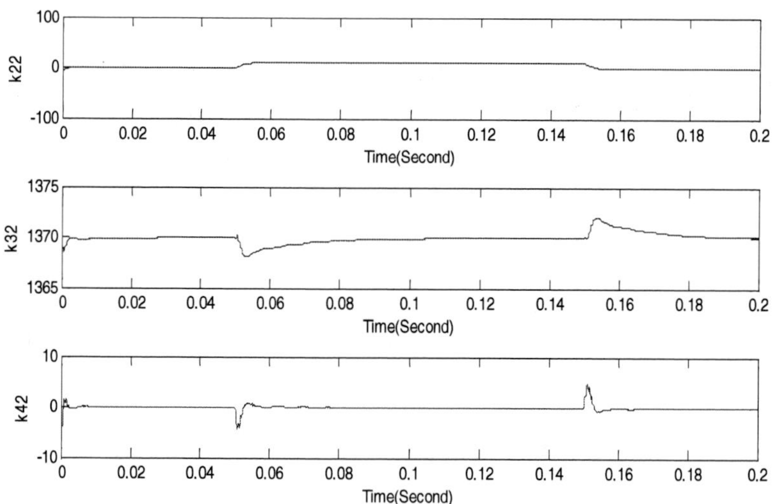

Figure 6.3. Adaptation of the controller parameters.

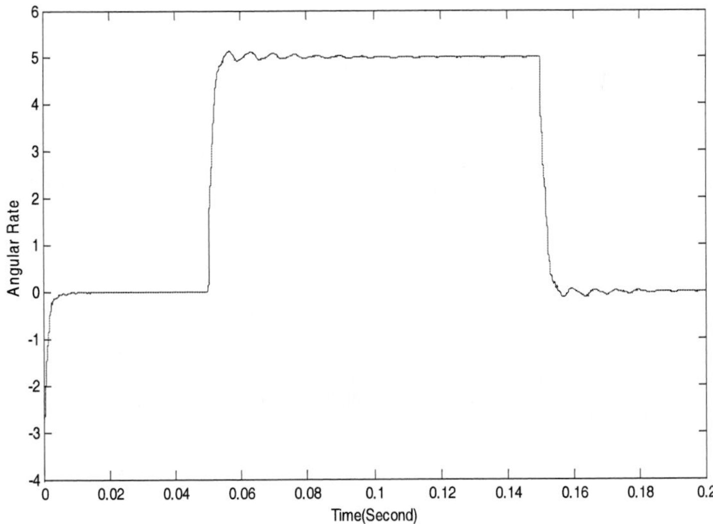

Figure 6.4. Adaptation of the angular velocity ($\Omega_z = 5.0$ rad/s).

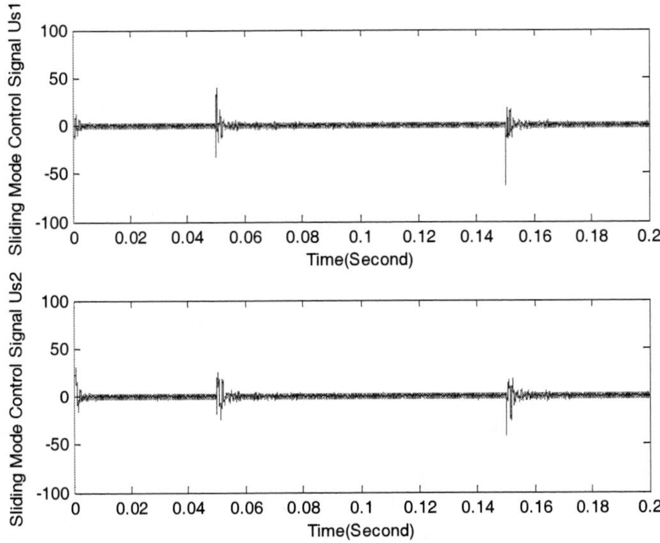

Figure 6.5. Smooth sliding mode control force of the adaptive sliding mode controller.

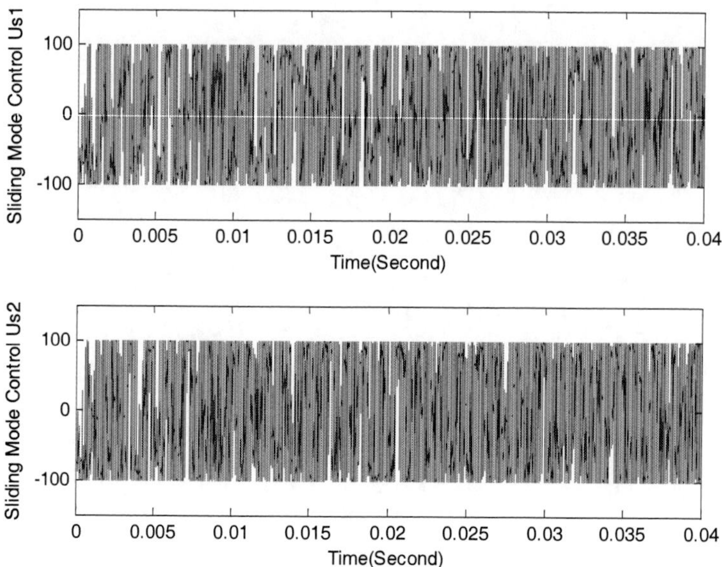

Figure 6.6. Non-smooth sliding mode control force of the adaptive sliding mode controller.

Figure 6.4 verifies that the estimate of angular velocity has good step tracking performance. Figure 6.5 plots the smooth sliding mode control force. It can be observed that the smooth adaptive sliding mode system significantly reduces chattering , as compared with the non-smooth sliding mode force as shown in Figure 6.6.

CONCLUDING REMARKS

This chapter investigated the design of a model reference adaptive state feedback control with a sliding mode property. A novel adaptive sliding mode controller with a proportional and integral sliding surface was proposed. The stability of closed-loop system could be established by the proposed adaptive sliding mode control structure with integral sliding action. The controller proposed here used a novel sliding mode algorithm, consisting of a proportional and integral sliding surface. An adaptive sliding mode controller was derived to control the axes of the gyroscope and to estimate the unknown angular velocity. Furthermore, a smooth version of the adaptive sliding mode controller was used to reduce the control chattering. The proposed adaptive sliding mode control structure, with proportional and integral sliding action, could handle both matched and unmatched uncertainties and disturbance; provided that upper bounds for these uncertainties were available. Simulation results demonstrated that the use of the proposed proportional-integral sliding mode adaptive control technique was effective in estimating the gyroscope parameters and angular velocity in the presence of matched and unmatched disturbances.

Chapter 7

ROBUST ADAPTIVE SLIDING MODE CONTROL OF MEMS GYROSCOPES

This chapter presents an adaptive sliding mode controller for a MEMS z-axis vibratory gyroscope. The adaptive model reference state tracking control with a sliding mode controller is developed to estimate the angular velocity and the damping and stiffness coefficients of MEMS vibratory gyroscope in real time. An adaptive controller that can estimate the unknown upper bound of the parameter uncertainties and external disturbance is derived and the stability of the closed-loop system is established. The adaptive sliding mode controller is designed so that the trajectory of the driving axes can track the state of the reference vibration model. The effectiveness of the proposed scheme is demonstrated through numerical simulations on a MEMS vibratory gyroscope.

This chapter incorporates the sliding mode control algorithm for the adaptive control system; and investigates the feasibility of adaptive sliding mode control for the MEMS angular velocity sensor, in the presence of the model uncertainty and external disturbance. The motivation of this chapter is to propose an adaptive sliding mode controller to estimate the angular velocity, all unknown gyroscope parameters, and unknown upper bounds of model uncertainties and external disturbances. This is accomplished by using an adaptive model reference controller in a unified control frame. The novelty of this chapter lies in that now two adaptive controllers, associated with sliding mode control, will have been successfully applied to the control of MEMS gyroscope, once in the presence of the model uncertainty and external disturbance with an unknown upper bound.

This chapter proposes a new adaptive sliding mode control strategy for MEMS gyroscopes. A detailed study of the sliding mode control algorithm is presented and solved in the presence of matched parameter uncertainties and external disturbances in the gyroscope model. The adaptive law to identify the unknown upper bound of the parameter uncertainties and external disturbance is also developed. The control target for the MEMS gyroscope is to design an adaptive sliding mode controller so that the trajectory of the driving axes can track the trajectory of reference model. The purpose of the sliding mode control is to establish conditions that the unknown angular velocity can be consistently estimated.

7.1. Robust Adaptive Sliding Mode Control

Consider the system in (2.28) with parametric uncertainties ΔA and external disturbance $f(t)$ as:

$$\dot{X}(t) = (A + \Delta A)X(t) + Bu + f(t) \tag{7.1}$$

where ΔA is unknown uncertainties of A, $f(t)$ is an uncertain disturbance or unknown nonlinearity of the system.

We make the following assumptions:

A1. There exist unknown matrices of appropriate dimensions D, G such that (i) $\Delta A(t) = BD(t)$, where $BD(t)$ is the matched uncertainty and (ii) $f(t) = BG(t)$, where $BG(t)$ is the matched disturbance. Therefore, (10) can be rewritten as:

$$\begin{aligned}\dot{X}(t) &= AX(t) + Bu(t) + \Delta AX(t) + f(t) \\ &= AX(t) + Bu(t) + BD(t)X(t) + BG(t) \\ &\equiv AX(t) + Bu(t) + Bf_m(t, X)\end{aligned} \tag{7.2}$$

where $Bf_m(t, X)$ represents the lumped, matched uncertainty and disturbance, given by $f_m(t, X) = DX(t) + G$.

A2. The matched lumped uncertainty and external disturbance f_m is bounded such as $\|f_m\| \leq \overline{\alpha}_1 + \overline{\alpha}_2\|X\|$, where $\overline{\alpha}_1$ and $\overline{\alpha}_2$ are unknown positive constants. Therefore, the proposed approach can be applied to bounded modeling uncertainty and disturbance.

A3. There exists a constant vector $K^* \in R^n$ such that the matching condition $A + BK^{*T} = A_m$ can be always satisfied , where A, A_m and B are constant matrix. The tracking error and its derivative are:

$$e(t) = X(t) - X_m(t) \tag{7.3}$$

$$\dot{e} = A_m e + (A - A_m)X + Bu + Bf_m. \tag{7.4}$$

The sliding surface is defined as $s(t) = \lambda e$, where λ is a constant matrix. The derivative of the sliding surface is:

$$\dot{s} = \lambda A_m e + \lambda(A - A_m)X + \lambda Bu + \lambda Bf_m. \tag{7.5}$$

Setting $\dot{s} = 0$ to solve equivalent control u_{eq} gives:

$$u_{eq} = K^{*T} X(t) - (\lambda B)^{-1} \lambda A_m e - f_m. \tag{7.6}$$

We consider the following sliding mode control law:

$$u(t) = -\gamma(\lambda B)^{-1} s + K^T(t)X(t) - (\lambda B)^{-1} \lambda A_m e(t) - \rho \frac{B^T \lambda^T s}{\|B^T \lambda^T s\|} \tag{7.7}$$

where γ is a positive constant, $K(t)$ is the estimate of K^*, $\rho = \alpha_1 + \alpha_2\|X\|$, $\|\cdot\|$ is the Euclidean norm, $u_s = \begin{bmatrix} u_{s1} \\ u_{s2} \end{bmatrix} = \frac{B^T \lambda^T s}{\|B^T \lambda^T s\|}$ is the sliding mode term representing the nonlinear feedback control for suppression of the effect of the uncertainty.

We define the estimation error as: $\widetilde{K}(t) = K(t) - K^*$, then we get

$$\dot{X}(t) = AX(t) + B\widetilde{K}^T(t)X(t) - \gamma B(\lambda B)^{-1}s + Bf_m - B\rho\frac{B^T\lambda^T s}{\left\|B^T\lambda^T s\right\|} \cdot \tag{7.8}$$

Subsequently, we have the tracking error equation:

$$\dot{e}(t) = (I - B(\lambda B)^{-1}\lambda)A_m e + B\widetilde{K}^T(t)X(t) - \gamma B(\lambda B)^{-1}s + Bf_m - B\rho\frac{B^T\lambda^T s}{\left\|B^T\lambda^T s\right\|} \cdot \tag{7.9}$$

The dynamics of sliding surface $s(t)$ is:

$$\dot{s}(t) = -\gamma s + \lambda B\widetilde{K}^T(t)X(t) + \lambda Bf_m - \rho\lambda B\frac{B^T\lambda^T s}{\left\|B^T\lambda^T s\right\|}. \tag{7.10}$$

Define a Lyapunov function:

$$V = \frac{1}{2}s^T s + \frac{1}{2}tr[\widetilde{K}M^{-1}\widetilde{K}^T] + \frac{1}{2}c_1\widetilde{\alpha}_1^{\ 2} + \frac{1}{2}c_2\widetilde{\alpha}_2^{\ 2} \tag{7.11}$$

where $M = M^T > 0$, $\widetilde{\alpha}_1 = \alpha_1 - \overline{\alpha}_1$, $\widetilde{\alpha}_2 = \alpha_2 - \overline{\alpha}_2$, M is a positive definite matrix.

Differentiating V , with respect to time, yields:

$$\begin{aligned}
\dot{V} &= s^T \dot{s} + tr[\tilde{K}M^{-1}\dot{\tilde{K}}^T] + c_1\tilde{\alpha}_1\dot{\tilde{a}}_1 + c_2\tilde{\alpha}_2\dot{\tilde{a}}_2 \\
&= s^T\left[-\gamma s + \lambda B\tilde{K}^T(t)X(t) + \lambda Bf_m - \rho\lambda B\frac{B^T\lambda^T s}{\left\|B^T\lambda^T s\right\|}\right] \\
&+tr[\tilde{K}M^{-1}\dot{\tilde{K}}^T] + c_1\tilde{\alpha}_1\dot{\tilde{a}}_1 + c_2\tilde{\alpha}_2\dot{\tilde{a}}_2 \\
&= -\gamma s^T s - \rho s^T\lambda B\frac{B^T\lambda^T s}{\left\|B^T\lambda^T s\right\|} + s^T\lambda Bf_m \\
&+tr[\tilde{K}B^T\lambda^T sX^T] + tr[\tilde{K}M^{-1}\dot{\tilde{K}}^T] + c_1\tilde{\alpha}_1\dot{\tilde{a}}_1 + c_2\tilde{\alpha}_2\dot{\tilde{a}}_2.
\end{aligned} \tag{7.12}$$

To make $\dot{V} \le 0$, we choose the adaptive laws as:

$$\dot{\tilde{K}}^T(t) = \dot{K}^T(t) = -MB^T\lambda^T sX^T(t) \tag{7.13}$$

with $K(0)$ being arbitrary. This adaptive law yields:

$$\begin{aligned}
\dot{V} &= -\gamma s^T s - \rho\left\|B^T\lambda^T s\right\| + s^T\lambda Bf_m + c_1\tilde{\alpha}_1\dot{\tilde{a}}_1 + c_2\tilde{\alpha}_2\dot{\tilde{a}}_2 \\
&= -\gamma s^T s - \left\|B^T\lambda^T s\right\|\left(\alpha_1 + \alpha_2\left\|X\right\|\right) + s^T\lambda Bf_m \\
&+c_1\left(\alpha_1 - \bar{\alpha}_1\right)\dot{\tilde{a}}_1 + c_2\left(\alpha_2 - \bar{\alpha}_2\right)\dot{\tilde{a}}_2 \\
&\le -\gamma s^T s - \left\|B^T\lambda^T s\right\|\left(\alpha_1 + \alpha_2\left\|X\right\|\right) + \left\|B^T\lambda^T s\right\|\left\|f_m\right\| \\
&+c_1\left(\alpha_1 - \bar{\alpha}_1\right)\dot{\tilde{a}}_1 + c_2\left(\alpha_2 - \bar{\alpha}_2\right)\dot{\tilde{a}}_2 \\
&\le -\gamma s^T s - \left\|B^T\lambda^T s\right\|\left(\alpha_1 + \alpha_2\left\|X\right\|\right) + \left\|B^T\lambda^T s\right\|\left(\bar{\alpha}_1 + \bar{\alpha}_2\left\|X\right\|\right) \\
&+c_1\left(\alpha_1 - \bar{\alpha}_1\right)\dot{\tilde{a}}_1 + c_2\left(\alpha_2 - \bar{\alpha}_2\right)\dot{\tilde{a}}_2 \\
&= -\gamma s^T s - \left(\bar{\alpha}_1 - \alpha_1\right)\left(\left\|B^T\lambda^T s\right\| - c_1\dot{\tilde{\alpha}}_1\right) \\
&+\left(\bar{\alpha}_2 - \alpha_2\right)\left(\left\|B^T\lambda^T s\right\|\left\|X\right\| - c_2\dot{\tilde{\alpha}}_2\right)
\end{aligned} \tag{7.14}$$

with the choice of the adaptive laws for the upper bound as:

$$\dot{\tilde{\alpha}}_1 = \frac{1}{c_1}\left\|B^T \lambda^T s\right\| \tag{7.15}$$

$$\dot{\tilde{\alpha}}_2 = \frac{1}{c_2}\left\|B^T \lambda^T s\right\|\left\|X\right\| . \tag{7.16}$$

This adaptive law yields $\dot{V} \le -\gamma s^T s \le 0$, $\dot{V}$ becomes negative semi-definite. This implies that s, $\tilde{K}$, $\tilde{\alpha}_1$ and $\tilde{\alpha}_2$ are all bounded. The property of $\dot{V} \le -\gamma s^T s$ follows that $V(t) - V(0) \le -\int_0^t \gamma s^T(\tau)s(\tau)d\tau$, i.e. $V(t) + \int_0^t \gamma s^T(\tau)s(\tau)d\tau \le V(0)$. The L_2-norm of s is bounded. Since $V(0)$ is bounded and $V(t)$ is nonincreasing and bounded, it can be concluded that $\lim_{t\to\infty}\int_0^t \|s\|dt$ is bounded. Since $\lim_{t\to\infty}\int_0^t \|s\|dt$ is bounded and $\dot{s}$ is also bounded, according to Barbalat lemma, $s(t)$ will asymptotically converge to zero, $\lim_{t\to\infty} s(t) = 0$, that is $s(t)$ and $e(t)$ all converge to zero asymptotically. This indicates that the sliding mode condition is satisfied and the robustness of stability can be guaranteed: $s(t)$ will asymptotically converge to zero, $\lim_{t\to\infty} s(t) = 0$, and then $e(t)$ also will asymptotically converge to zero. This indicates that the sliding mode condition is satisfied and the robustness of stability can be guaranteed. The inequality $\dot{V} \le -\eta\|s\|$ implies that s is integrable as $\int_0^t \|s\|dt \le \frac{1}{\eta}\left[V(0) - V(t)\right]$. Since $V(0)$ is bounded and $V(t)$ is nonincreasing and bounded, it can be concluded that $\lim_{t\to\infty}\int_0^t \|s\|dt$ is bounded. Since $\lim_{t\to\infty}\int_0^t \|s\|dt$ is bounded and $\dot{s}$ is also bounded, according to Barbalat lemma, $s(t)$ will asymptotically converge to zero, $\lim_{t\to\infty} s(t) = 0$, that is $s(t)$ and $e(t)$ all converge to zero asymptotically. If X is a persistent excitation signal, then $\dot{\tilde{K}}^T(t) = -MB^T\lambda^T sX^T$ guarantees that

$\tilde{K} \to 0$. It can be shown that if $w_1 \neq w_2$, there always exists some positive scalar constants α and T , such that for all $t > 0$, $\int_t^{t+T} XX^T d\tau \geq \alpha I$,

where $XX^T = \begin{bmatrix} x_1^2 & x_1\dot{x}_1 & x_1x_2 & x_1\dot{x}_2 \\ \dot{x}_1x_1 & \dot{x}_1^2 & \dot{x}_1x_2 & \dot{x}_1\dot{x}_2 \\ x_2x_1 & x_2\dot{x}_1 & x_2^2 & x_2\dot{x}_2 \\ \dot{x}_2x_1 & \dot{x}_2\dot{x}_1 & \dot{x}_2x_2 & \dot{x}_2^2 \end{bmatrix}$.

It can be shown that XX^T has full rank if $w_1 \neq w_2$, i.e. the excitation frequencies on x and y axes should be different. In other words, excitation of proof mass should be persistently exciting. Since $\tilde{K} \to 0$, then the unknown angular velocity as well as all other unknown parameters can be determined from $A + BK^T = A_m$.

In summary, if persistently exciting drive signals, $x_m = A_1 \sin(w_1 t)$ and $y_m = A_2 \sin(w_2 t)$ are used, then $\tilde{K}$, $s(t)$ and $e(t)$ all converge to zero asymptotically. Consequently, the unknown angular velocity can be determined as $\lim_{t\to\infty} \hat{\Omega}_z(t) = \Omega_z$. It is difficult, however to establish the convergence rate.

Remark 1. In order to eliminate chattering, the discontinuous control component in (7.7) can be replaced by a smooth sliding mode component to yield:

$$u(t) = -\gamma(\lambda B)^{-1}s + K^T(t)X(t) + K_e e - \rho\lambda B\frac{B^T\lambda^T s}{\left\|B^T\lambda^T s\right\| + \varepsilon} \tag{7.17}$$

where $\varepsilon > 0$ is a small constant. This creates a small boundary layer around the switching surface in which the trajectory will remain. Therefore, chattering can be reduced.

Remark 2. At times, s will not always be equal to zero if the continuous approximation for the sliding mode law is used; therefore, the adaptive gains will slowly increase boundlessly. This will decrease tracking the accuracy of the control system. The dead-zone techniques can be used to remove this implementation problem. To eliminate the problem of integral wind-up in the adaptation of the upper bound of the unknown disturbance, the adaptive laws are modified as:

$$\dot{\tilde{\alpha}}_1 = \frac{1}{c_1}\left(-\varphi_1\tilde{\alpha}_1 + \left\|B^T\lambda^T s\right\|\right), \tag{7.18}$$

$$\dot{\tilde{\alpha}}_2 = \frac{1}{c_2}\left(-\varphi_2\tilde{\alpha}_2 + \left\|B^T\lambda^T s\right\|\left\|X\right\|\right) \tag{7.19}$$

where φ_1 and φ_2 are constant.

7.2. Simulation Example

In order to verify the validity of the proposed controller, we implement the controller on the lumped MEMS gyroscope model.

The control objective is to design an adaptive sliding mode controller so that the trajectory of $X(t)$ can track the reference model $X_m(t)$.

In the simulation, we allowed $\pm$ 10% parameter variations for the spring and damping coefficients; further assuming $\pm$ 5% magnitude changes in the coupling terms i.e. d_{xy} and ω_{xy}.

All the above parameter variations are included in the uncertainties of the system parameters ΔA. The external disturbance is a random variable with zero mean and unity variance. Parameters of the MEMS gyroscope are as follows:

$$m = 0.57\times10^{-8}\text{ kg},\ d_{xx} = 0.429\times10^{-6}\text{ N s/m},$$

$$d_{xy} = 0.0429\times10^{-6}\text{ N s/m},\ d_{yy} = 0.687\times10^{-6}\text{ N s/m}$$

$$k_{xx} = 80.98\text{ N/m},\ k_{xy} = 5\text{ N/m},\ k_{yy} = 71.62\text{ N/m}$$

$$w_0 = 3kHz,\ q_0 = 10^{-6}m.$$

The unknown angular velocity Ω_z is assumed to be a step function of amplitude 5.0 rad/s and the initial condition on K matrix is $K(0) = 0.9K^*$. The desired motion trajectories are $x_m = \sin(w_1 t)$ and $y_m = 1.2\sin(w_2 t)$, where $w_1 = 4.11kHz$ and $w_2 = 5.11kHz$.

The sliding mode parameter matrix λ is chosen as $\lambda = \begin{bmatrix} 1 & 10 & 0 & 10 \\ 10 & 0 & 10 & 10 \end{bmatrix}$, $M = diag\{5 \quad 5\}$, $\varepsilon = 0.05$, $c_1 = c_2 = 20$, $\varphi_1 = \varphi_2 = 10$, $\gamma = 1000$.

The behavior of tracking error and sliding surface are shown in Figures 7.1-7.3. Figure 7.4 verifies that the estimation of angular velocity has satisfactory step tracking performance. Figure 7.5 shows that adaptation of the upper bound of the disturbance's magnitude. Figure 7.7 demonstrates the adaptation of controller parameters. Figure 7.7 plots the sliding mode control force for the smooth sliding mode controller.

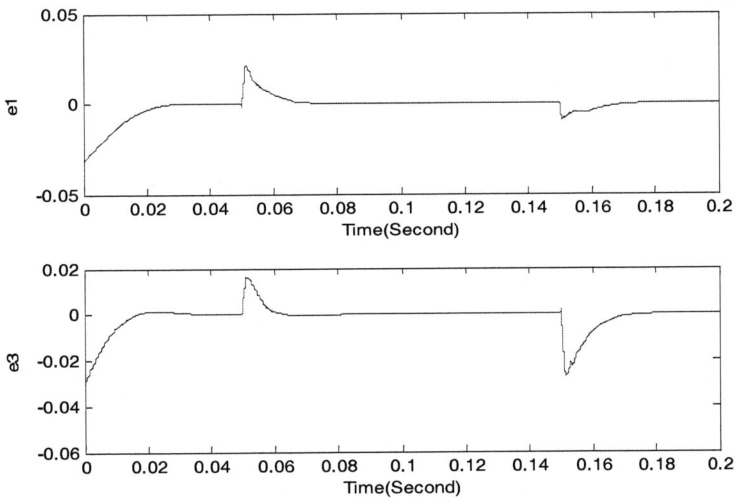

Figure 7.1. Plot of the tracking error e(t).

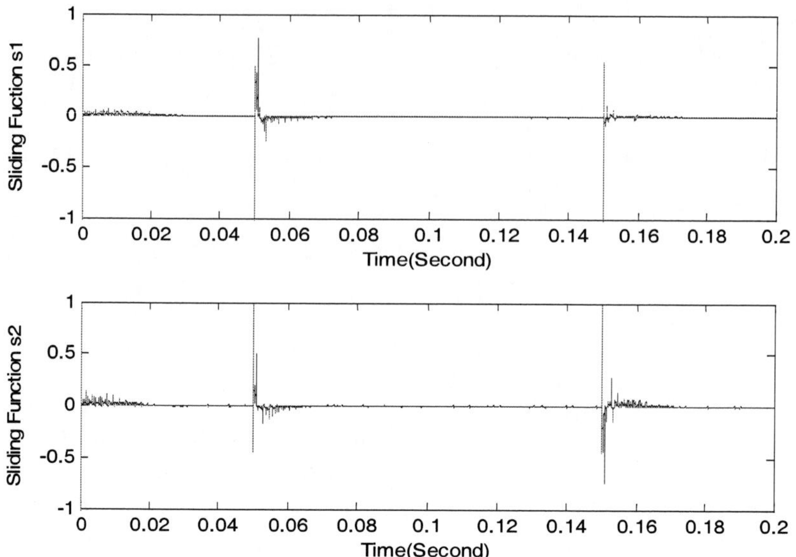

Figure 7.2. Plot of the sliding surface s(t).

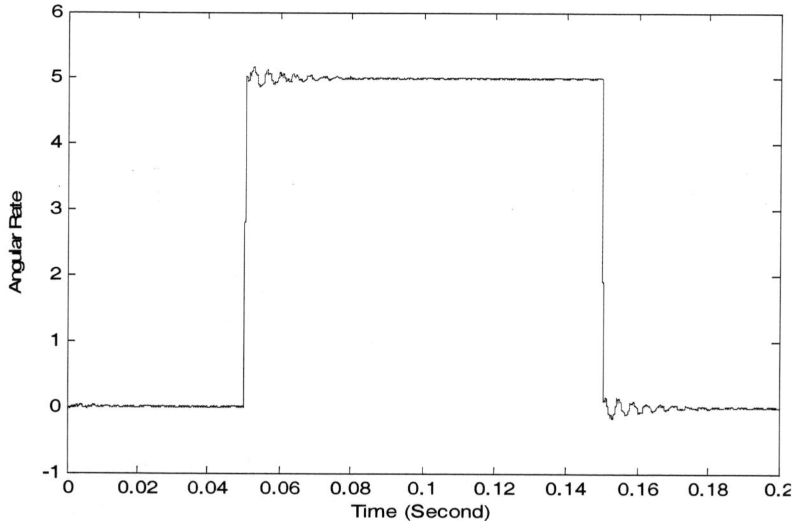

Figure 7.3. Adaptation of angular velocity.

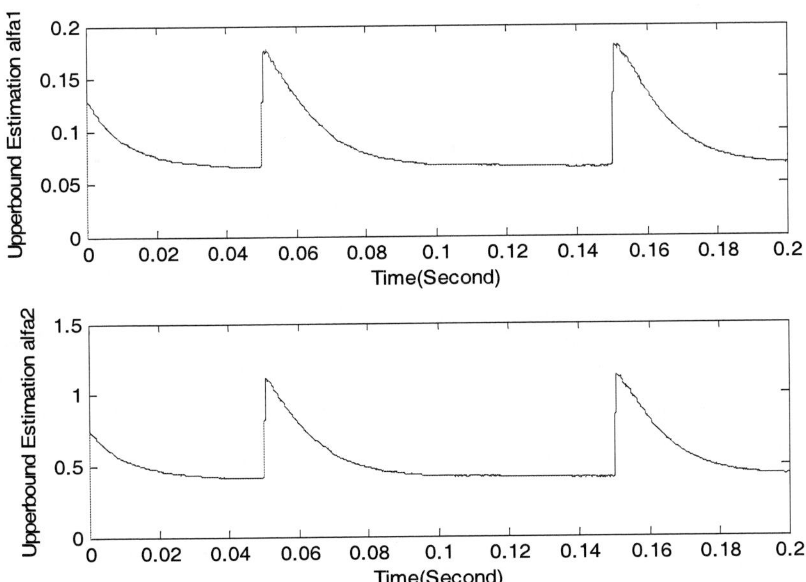

Figure 7.4. Adaptation of upper bound of disturbance.

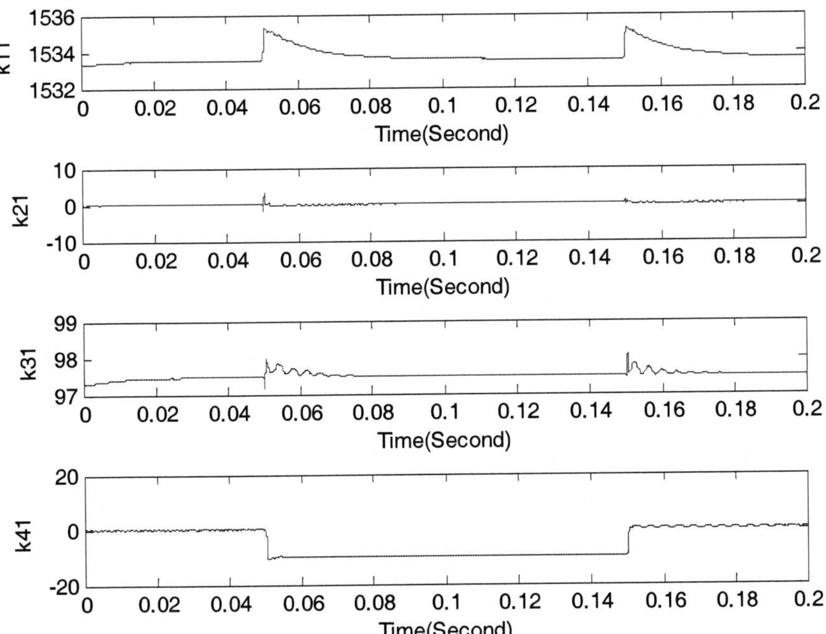

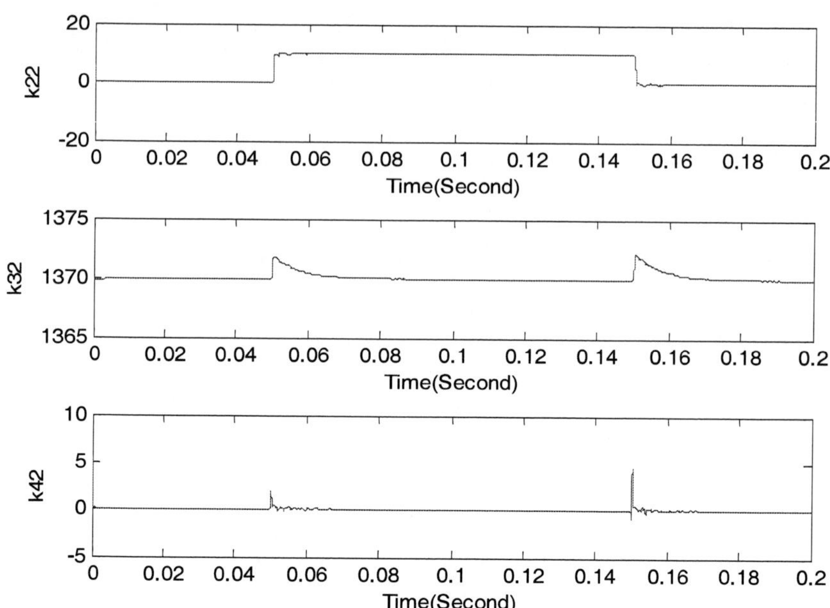

Figure 7.5. Adaptation of controller parameters.

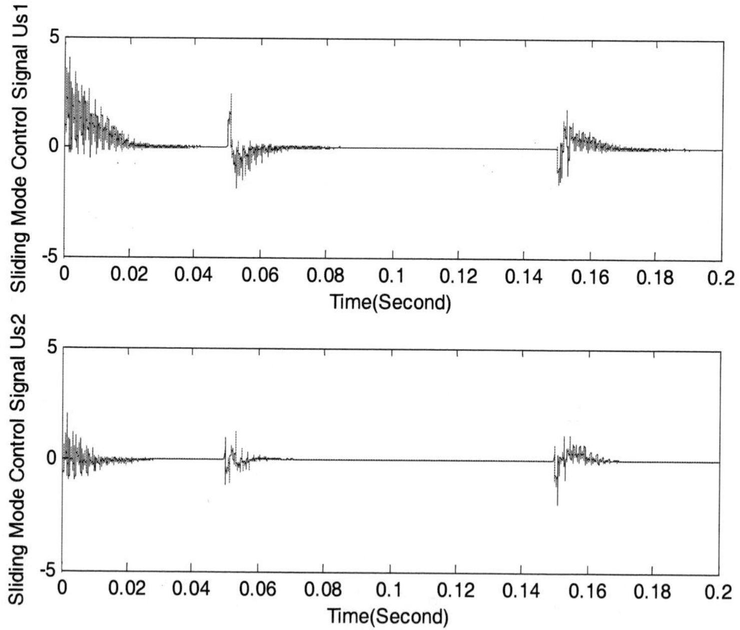

Figure 7.6. Smooth sliding mode control force of the adaptive sliding mode controller.

CONCLUDING REMARKS

This chapter investigated the design of a model reference adaptive state feedback sliding mode control for MEMS angular velocity sensor. The controller proposed here uses a standard sliding mode algorithm. An adaptive sliding mode controller was derived to control the axes of the gyroscope and to estimate the unknown angular velocity. The adaptive law to estimate the unknown upper bound of the parameter uncertainties and external disturbances was also derived. Furthermore, a smooth version of the adaptive sliding mode controller was used to reduce the control chattering. The proposed adaptive sliding mode control structure could establish the stability of a closed-loop system. The simulation results demonstrated that the use of the proposed sliding mode control technique was effective in estimating the gyroscope parameters and angular velocity in the presence of external disturbance and model uncertainties.

Chapter 8

ADAPTIVE SLIDING MODE CONTROL AND OBSERVER OF MEMS GYROSCOPES

This chapter presents a novel adaptive sliding mode control with a sliding mode observer for a MEMS gyroscope. The proposed adaptive sliding mode controller with a sliding mode observer can estimate the unmeasured velocities, angular velocity and damping and stiffness coefficients of the gyroscope in the presence of parameter variations and external disturbance. A novel adaptive sliding mode controller with a proportional and integral sliding surface is derived and the stability condition of the closed-loop feedback system is established. Simulation results demonstrate the effectiveness of the proposed adaptive sliding mode controller with sliding mode observer.

The basic sliding mode observer structure consists of switching terms added to a conventional Luenberger observer. The sliding mode design method enhances the robustness over a range of system uncertainties and disturbances. The motivation of this chapter is to combine a sliding mode observer with a novel adaptive sliding mode control for a MEMS gyroscope. The chapter presents a novel adaptive sliding mode controller, in conjunction with a sliding mode observer which can reconstruct estimates of the unmeasured velocity states. An adaptive sliding mode controller with a proportional and integral sliding surface is derived and the closed-loop stability of the adaptive sliding mode controller with a sliding mode observer is established. The proposed adaptive sliding mode controller can estimate the angular velocity and all the gyroscope parameters in real time in the presence of unmeasured states.

8.1. Sliding Mode Controller and Sliding Mode Observer Design

In this section, an algorithm for the sliding mode control and sliding mode observer will be developed. Consider the system in (2.28) with parametric uncertainties ΔA and external disturbance $f(t)$ as:

$$\dot{X}(t) = (A + \Delta A)X(t) + Bu + f(t) \tag{8.1}$$

where ΔA is unknown uncertainties of the matrix A, $f(t)$ is an uncertain disturbance or unknown nonlinearity of the system.

We assume that not all the states $X = [x \;\; \dot{x} \;\; y \;\; \dot{y}]^T$ are measurable; only position signals x and y are measurable; therefore, the system output becomes:

$$Y(t) = CX(t) \tag{8.2}$$

where $C = \begin{bmatrix} 1 & 0 & 0 & 0 \\ 0 & 0 & 1 & 0 \end{bmatrix}$. (8.3)

The control target for MEMS gyroscope is maintaining the oscillation of the proof mass in x and y directions, at frequencies w_1, w_2 ; at amplitudes A_1 , A_2 , respectively. These requirements can be expressed as: $x_m = A_1 \sin(w_1 t)$, and $y_m = A_2 \sin(w_2 t)$.

We make the following assumptions concerning uncertainties:

A1. There exist unknown matrices of appropriate dimensions D, G such that (i) $\Delta A(t) = BD(t)$, where $BD(t)$ is the matched uncertainty and (ii) $f(t) = BG(t)$, where $BG(t)$ is the matched disturbance. Therefore, (8.1) can be rewritten as:

$$\dot{X}(t) = AX(t) + Bu(t) + \Delta AX(t) + f(t)$$
$$= AX(t) + Bu(t) + BD(t)X(t) + BG(t) \quad (8.4)$$
$$\equiv AX(t) + Bu(t) + Bf_m(t, X)$$

where $Bf_m(t, X)$ represents the lumped matched uncertainty and disturbance which is given by:

$$f_m(t, X) = DX(t) + G. \quad (8.5)$$

A2. The matched uncertainty $f_m(t, X)$ is bounded, such as: $\|f_m(t, X)\| \le \alpha_{m1}\|X\| + \alpha_{m2}$ where α_{m1}, α_{m2} are known positive constants, satisfying $\alpha_{m1}\|X\| << \alpha_{m2}$. In chapter 7, we consider a nonzero unmatched disturbance in connection with a sliding mode control that does not use an observer.

A3. There exists a constant matrix K^* such that the following matching condition $A + BK^{*T} = A_m$ can always be satisfied.

Remark 1. Solving $A + BK^{*T} = A_m$ yields $K^{*T} = (B^T B)^{-1} B^T (A_m - A)$, i.e.,

$$\begin{bmatrix} k_{11}^* & k_{21}^* & k_{31}^* & k_{41}^* \\ k_{12}^* & k_{22}^* & k_{32}^* & k_{42}^* \end{bmatrix} = \begin{bmatrix} -w_1^2 + w_x^2 & d_{xx} & w_{xy} & d_{xy} - 2\Omega_z \\ w_{xy} & d_{xy} + 2\Omega_z & -w_2^2 + w_y^2 & d_{yy} \end{bmatrix}$$

Therefore, the solution can be obtained as $w_x{}^2 = k_{11}{}^* + w_1{}^2$, $w_y{}^2 = k_{32}{}^* + w_2{}^2$, $d_{xx} = k_{21}{}^*$, $w_{xy} = k_{31}{}^* = k_{12}{}^*$ and $d_{yy} = k_{42}{}^*$. From $d_{xy} + 2\Omega_z = k_{22}{}^*$ and $d_{xy} - 2\Omega_z = k_{41}{}^*$, $d_{xy} = \frac{1}{2}(k_{22}{}^* + k_{41}{}^*)$ and $\Omega_z = \frac{1}{4}(k_{22}{}^* - k_{41}{}^*)$ can be obtained.

It can be seen that the unknown gyroscope parameters such as d_{xx}, d_{xy}, d_{yy}, w_x^2, w_{xy}, w_y^2 and angular velocity Ω_z can be determined by the controller parameters K^*.

Remark 2. The limitation of direct adaptive sliding mode controller is that controller structures have only eight components which can only correspond to at most eight system parameters. The gyroscope parameters are determined by controller parameters indirectly. In order to obtain the unique gyroscope system parameter solution, some restrictions have to be imposed on the controller structure such as $k_{31} = k_{12}$.

The tracking error and its derivative are:

$$e(t) = X(t) - X_m(t) \tag{8.6}$$

$$\dot{e} = A_m e + (A - A_m)X + Bu + Bf_m. \tag{8.7}$$

The proportional and integral sliding surface $s(t) = 0$ is defined as:

$$s(t) = \lambda e - \int_0^t \lambda(A_m + BK_e)ed\tau \tag{8.8}$$

where λ is a constant matrix, $\lambda = \begin{bmatrix} \lambda_{11} & \lambda_{12} & \lambda_{13} & \lambda_{14} \\ \lambda_{21} & \lambda_{22} & \lambda_{23} & \lambda_{24} \end{bmatrix}$ which satisfies the condition that λB is nonsingular. The constant K_e satisfies the condition that $(A_m + BK_e)$ is Hurwitz.

Setting $\dot{s} = 0$ gives the equivalent control u_{eq}

$$\dot{s} = \lambda(A - A_m)X + \lambda Bu + \lambda Bf_m - \lambda BK_e e \tag{8.9}$$

$$\begin{aligned} u_{eq} &= -(\lambda B)^{-1}\lambda(A - A_m)X + K_e e - f_m \\ &= K^{*T}X(t) + K_e e - f_m. \end{aligned} \tag{8.10}$$

Substituting u_{eq} into (8.7) provides the error dynamics:

$$\dot{e}(t) = (A_m + BK_e)e(t). \tag{8.11}$$

The control signal u is proposed as:

$$u = K^{*T} X(t) + K_e e - \rho_1 (\lambda B)^{-1} \frac{s}{\|s\|} \tag{8.12}$$

where $u = \begin{bmatrix} u_1 \\ u_2 \end{bmatrix}$, $K^{*T} = \begin{bmatrix} k_{11}^* & k_{21}^* & k_{31}^* & k_{41}^* \\ k_{12}^* & k_{22}^* & k_{32}^* & k_{42}^* \end{bmatrix}$. The last component of the control signal u is designed to address the matching model uncertainty and disturbance. This component is given as $u_s = \begin{bmatrix} u_{s1} \\ u_{s2} \end{bmatrix} = -\rho_1 (\lambda B)^{-1} \frac{s}{\|s\|}$, where ρ_1 is a constant.

In practical MEMS gyroscopes, not all the state variables are available. In order to extend the practical applicability of the proposed control scheme, we can utilize the sliding mode observer to estimate the unmeasured states. Since the sliding mode observer has many advantages over the Luenberger observers such as robustness to parameter uncertainty, a sliding mode observer will be incorporated to provide estimates of the unmeasured states. Let $\hat{X}$ be the estimate of X, the observer error can be defined as:

$$\widetilde{X} = \hat{X} - X = \begin{bmatrix} \widetilde{x}_1 \\ \widetilde{x}_2 \\ \widetilde{x}_3 \\ \widetilde{x}_4 \end{bmatrix} = \begin{bmatrix} \hat{x}_1 - x_1 \\ \hat{x}_2 - x_2 \\ \hat{x}_3 - x_3 \\ \hat{x}_4 - x_4 \end{bmatrix}. \tag{8.13}$$

Consider the sliding mode observer structure which is given by:

$$\begin{aligned} \dot{\hat{X}}(t) &= A\hat{X}(t) + Bu - L(C\hat{X} - Y) + Bv_0 \\ &= A\hat{X}(t) + Bu - LC\widetilde{X} + Bv_0. \end{aligned} \tag{8.14}$$

The discontinuous vector v_0 is:

$$v_0 = \begin{cases} -\rho_2 \dfrac{FC\widetilde{X}}{\left\|FC\widetilde{X}\right\|} & \textit{if } FC\widetilde{X} \neq 0 \\ 0 & \textit{otherwise} \end{cases} \tag{8.15}$$

where ρ_2 is a scalar.

If the system is observable, the observer gain matrix L can be chosen so that the closed-loop matrix $(A - LC)$ is stable and has a Lyapunov matrix $P = P^T > 0$ satisfying:

(i) $(A - LC)^T P + P(A - LC) = -Q$ (8.16)

for some positive definite design matrix $Q = Q^T > 0$ and,

(ii) the structural constraint $FC = B^T P$, for some non-singular matrix F.
The observer error dynamics becomes:

$$\dot{\widetilde{X}} = (A - LC)\widetilde{X} + Bv_0 - Bf_m. \tag{8.17}$$

For the gyroscope dynamics (2.28) and with the choice of $L = \begin{bmatrix} l_{11} & l_{12} & l_{13} & l_{14} \\ l_{21} & l_{22} & l_{23} & l_{24} \end{bmatrix}^T$ and $F = \begin{bmatrix} 0.707 & 0.707 \\ 0.707 & 0.707 \end{bmatrix}^T$, the sliding mode observer can be constructed as:

$$\begin{aligned}
\dot{\hat{x}}_1 &= \hat{x}_2 - l_1\widetilde{x}_1 \\
\dot{\hat{x}}_2 &= -w_x^2\hat{x}_1 - d_{xx}\hat{x}_2 - w_{xy}\hat{x}_3 - (d_{xy} - 2\Omega_z)\hat{x}_4 - l_2\widetilde{x}_1 - \rho_2 \operatorname{sgn}(\widetilde{x}_1 + \widetilde{x}_3) + u_x \\
\dot{\hat{x}}_3 &= \hat{x}_4 - l_3\widetilde{x}_1 \\
\dot{\hat{x}}_4 &= -w_{xy}\hat{x}_1 - (d_{xy} + 2\Omega_z)\hat{x}_2 - w_x^2\hat{x}_3 - d_{yy}\hat{x}_4 - l_4\widetilde{x}_1 - \rho_2 \operatorname{sgn}(\widetilde{x}_1 + \widetilde{x}_3) + u_y.
\end{aligned} \tag{8.18}$$

The observer error dynamics becomes:

$$
\begin{aligned}
\dot{\tilde{x}}_1 &= \tilde{x}_2 - l_{11}\tilde{x}_1 - l_{21}\tilde{x}_3 \\
\dot{\tilde{x}}_2 &= -w_x^2\tilde{x}_1 - d_{xx}\tilde{x}_2 - w_{xy}\tilde{x}_3 - (d_{xy} - 2\Omega_z)\tilde{x}_4 - l_{12}\tilde{x}_1 - l_{22}\tilde{x}_3 - \rho_2 \operatorname{sgn}(\tilde{x}_1 + \tilde{x}_3) - f_{m1} \\
\dot{\tilde{x}}_3 &= \tilde{x}_4 - l_{13}\tilde{x}_1 - l_{23}\tilde{x}_3 \\
\dot{\tilde{x}}_4 &= -w_{xy}\tilde{x}_1 - (d_{xy} + 2\Omega_z)\tilde{x}_2 - w_x^2\tilde{x}_3 - d_{yy}\tilde{x}_4 - l_{14}\tilde{x}_1 - l_{24}\tilde{x}_3 - \rho_2 \operatorname{sgn}(\tilde{x}_1 + \tilde{x}_3) - f_{m2}
\end{aligned}
\tag{8.19}
$$

where $f_m = \begin{bmatrix} f_{m1} & f_{m2} \end{bmatrix}^T$.

8.2. Adaptive Sliding Mode Controller Design

In this section, the effect of using the state estimates in the control law will be investigated. Both states, angular velocity Ω_z and gyroscope parameters will be estimated in real time using an adaptive sliding mode controller with a sliding mode observer. It will be shown that the control performance is robust to estimation errors when used in conjunction with the sliding mode observer. The block diagram of the adaptive sliding mode control with sliding mode observer is shown in Figure 8.1.

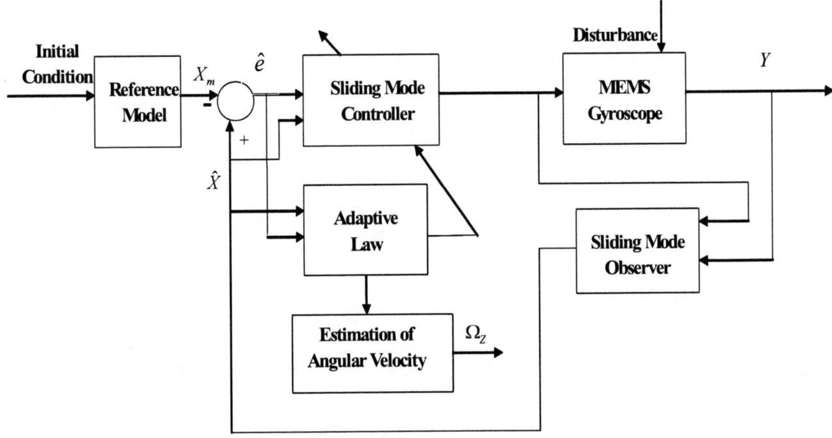

Figure 8.1. Block diagram of adaptive sliding mode control with sliding mode observer.

We define the tracking error between the state estimates and the reference model state as:

$$\hat{e}(t) = \hat{X}(t) - X_m(t) . \tag{8.20}$$

The sliding surface in terms of $\hat{e}(t)$ is defined as:

$$\hat{s}(t) = \lambda \hat{e} - \int_0^t \lambda (A_m + BK_e)\hat{e} d\tau . \tag{8.21}$$

An adaptive version of control input $u(t)$ with the state estimates is proposed as:

$$u(t) = K^T(t)\hat{X}(t) + K_e \hat{e} + (\lambda B)^{-1} v_c . \tag{8.22}$$

Note that $u(t)$ in (8.22) share the same structure with $u(t)$ in (8.12), the latter corresponds to the known states $X(t)$. $K^T(t) = \begin{bmatrix} k_{11} & k_{21} & k_{31} & k_{41} \\ k_{12} & k_{22} & k_{32} & k_{42} \end{bmatrix}$ is an estimate of K^* in (8.12).

The robust component of the control signal is defined by:

$$v_c = \begin{cases} -\rho_1 \dfrac{\hat{s}}{\|\hat{s}\|} & if\ s \neq 0 \\ 0 & otherwise. \end{cases} \tag{8.23}$$

We define the controller gain estimation error as:

$$\widetilde{K}(t) = K(t) - K^* . \tag{8.24}$$

Substituting (8.24) and (8.22) into (8.14) yields:

$$\dot{\hat{X}}(t) = A_m \hat{X}(t) + B\widetilde{K}^T(t)\hat{X}(t) + BK_e(\hat{X} - X_m) + B(\lambda B)^{-1} v_c - LC\widetilde{X} + Bv_0 . \tag{8.25}$$

Using (8.20) and (8.25), the derivative of $\hat{e}(t)$ is obtained as:

$$\dot{\hat{e}}(t) = (A_m + BK_e)\hat{e} + B\widetilde{K}^T(t)\hat{X}(t) + B(\lambda B)^{-1}v_c - LC\widetilde{X} + Bv_0. \tag{8.26}$$

The sliding surface dynamics in terms of the state estimates becomes

$$\dot{\hat{s}}(t) = \lambda B\widetilde{K}^T(t)\hat{X}(t) + v_c - \lambda LC\widetilde{X} + \lambda Bv_0. \tag{8.27}$$

In order to show the stability of the closed-loop system, we define a Lyapunov function:

$$V = \frac{1}{2}\hat{s}^T\hat{s} + \frac{1}{2}tr[\widetilde{K}M^{-1}\widetilde{K}^T] + \frac{1}{2}\widetilde{X}^T P\widetilde{X} \tag{8.28}$$

where the P and $M = diag\{m_1 \quad m_2\}$ are positive definite matrices.

Differentiating V with respect to time yields:

$$\dot{V} = -\rho_1\|\hat{s}\| - \hat{s}^T\lambda LC\widetilde{X} + \hat{s}^T\lambda Bv_0 + \hat{s}^T\lambda B\widetilde{K}^T(t)\hat{X}(t) + tr[\widetilde{K}M^{-1}\dot{\widetilde{K}}^T] - \frac{1}{2}\widetilde{X}^T Q\widetilde{X} \\ -\rho_2\frac{\widetilde{X}^T PBFC\widetilde{X}}{\|FC\widetilde{X}\|} - \widetilde{X}^T PBf_m. \tag{8.29}$$

Using (8.18) and $FC = B^T P$ yields:

$$\dot{V} = -\rho_1\|\hat{s}\| - \hat{s}^T\lambda LC\widetilde{X} + \hat{s}^T\lambda Bv_0 - \frac{1}{2}\widetilde{X}^T Q\widetilde{X} - \rho_2\|FC\widetilde{X}\| - \widetilde{X}^T C^T F^T f_m \\ + \left(\hat{s}^T\lambda B\widetilde{K}^T\hat{X} + tr[\widetilde{K}M^{-1}\dot{\widetilde{K}}^T]\right). \tag{8.30}$$

In order to make $\dot{V}$ negative definite, the updating law is chosen as:

$$\dot{\widetilde{K}}^T(t) = \dot{\hat{K}}^T(t) = -MB^T\lambda^T\hat{s}\hat{X}^T(t) \tag{8.31}$$

with $\hat{K}(0)$ being arbitrary. This choice yields:

$$\begin{aligned}
\dot{V} &= -\rho_1 \|\hat{s}\| - \hat{s}^T \lambda LC\tilde{X} + \hat{s}^T \lambda Bv_0 - \frac{1}{2}\tilde{X}^T Q\tilde{X} \\
&-\rho_2 \|FC\tilde{X}\| - \tilde{X}^T C^T F^T f_m \\
&\leq -\rho_1 \|\hat{s}\| + \|\hat{s}\| \|\lambda L\| \|C\tilde{X}\| + \|\hat{s}\| \|\lambda B\| \rho_2 - \frac{1}{2}\tilde{X}^T Q\tilde{X} \\
&-\rho_2 \|FC\tilde{X}\| + \|FC\tilde{X}\| \|f_m\| \qquad (8.32) \\
&\leq -\|\hat{s}\| \left(\rho_1 - \|\lambda L\| \|\tilde{x}_1 + \tilde{x}_3\| - \rho_2 \|\lambda B\|\right) - \frac{1}{2}\tilde{X}^T Q\tilde{X} \\
&-\|FC\tilde{X}\| \left(\rho_2 - \alpha_{m1} \|X\| - \alpha_{m2}\right).
\end{aligned}$$

with the choices of $\rho_2 > \alpha_{m1}\|X\| + \alpha_{m2}$ and $\rho_1 > \|\lambda L\| \|\tilde{x}_1 + \tilde{x}_3\| + \rho_2 \|\lambda B\|$, $\dot{V}$ becomes negative semi-definite, i.e., $\dot{V} \leq 0$. This implies that $\hat{s}$, $\tilde{K}$, and $\tilde{X}$ are all bounded. $\dot{V}$ is negative definite implies that $\hat{s}$, $\tilde{K}$ and $\tilde{X}$ all converge to zero. The property of $\dot{V}$ is negative semi-definite ensures that V, $\hat{s}$, $\tilde{K}$ and $\tilde{X}$ are all bounded. $\dot{V} = 0$ implies $\tilde{X}(t) = 0$ and $\hat{s} = 0$. LaSalle's invariant set theorem can be used to prove that $\lim_{t\to\infty} \hat{s}(t) = 0$. $\dot{V} = 0$ implies that $\hat{s} = 0$ and there is no other solution but $\hat{s} = 0$. According to LaSalle's invariant set theorem and defining $R = \{\hat{s} \in R^n \quad \dot{V}(x) = 0\}$, then if R contains no other trajectories other than $\hat{s} = 0$, the origin 0 is asymptotically stable. Consequently the sliding surface $\hat{s} = 0$ is an invariant set which implies that any trajectory starting from an initial condition within the set remains in the set all the time, that is $\hat{s}(t)$ will asymptotically converge to zero, $\lim_{t\to\infty} \hat{s}(t) = 0$.

In order to prove that the parameter errors $\tilde{K}$ converge to zero, we need to make the persistence of excitation argument. From the adaptive law $\dot{\tilde{K}}^T(t) = -MB^T \lambda^T \hat{s}\hat{X}^T(t)$, and according to the persistent excitation theory, if $\hat{X}$ meets persistence of excitation condition, then $K(t)$ will converges to its true value. It can be shown that there exist some positive scalar constants α and T such that for all $t > 0$, $\int_{t}^{t+T} \hat{X}\hat{X}^{T} d\tau \geq \alpha I$,

where $\hat{X}\hat{X}^{T} = \begin{bmatrix} x_1^{2} & x_1\dot{x}_1 & x_1x_2 & x_1\dot{x}_2 \\ \dot{x}_1x_1 & \dot{x}_1^{2} & \dot{x}_1x_2 & \dot{x}_1\dot{x}_2 \\ x_2x_1 & x_2\dot{x}_1 & x_2^{2} & x_2\dot{x}_2 \\ \dot{x}_2x_1 & \dot{x}_2\dot{x}_1 & \dot{x}_2x_2 & \dot{x}_2^{2} \end{bmatrix}$. If $w_1 \neq w_2$, i.e. the excitation frequencies on x and y axes should be different it can be shown that $\hat{X}\hat{X}^{T}$ has full rank. Since $\hat{X}$ meets persistence of excitation condition, then $\tilde{K} \to 0$, i.e. the unknown angular velocity as well as all other unknown parameters can be consistently estimated. In consequence, angular velocity Ω_z and gyroscope parameters converge to their true values.

Conclusion: If persistently exciting drive signals, $x_m = A_1 \sin(w_1 t)$ and $y_m = A_2 \sin(w_2 t)$ are used, then $\tilde{K}(t)$, $\tilde{X}(t)$, $\hat{s}(t)$ and $\hat{e}(t)$ all asymptotically converge to zero. Consequently the unknown angular velocity can be determined as $\lim_{t\to\infty} \hat{\Omega}_z(t) = \Omega_z$. However it is difficult to establish the convergence rate.

Remark 1. It is not easy to guarantee that the stated inequality condition on ρ_2 can always be satisfied for all times and all state. However if $\alpha_{m1} << \alpha_{m2}$, then the condition becomes state independent and knowing α_{m2} can guarantee inequality on ρ_2. It is also critical to have ρ_2 as small as possible in order to avoid saturation in actuators.

Remark 2. In order to eliminate the chattering, the discontinuous control component $(\lambda B)^{-1} v_c$ in (8.22) can be replaced by a smooth sliding mode component to yield:

$$u(t) = K^{T}(t)\hat{X}(t) + K_e \hat{e} - (\lambda B)^{-1} \rho_1 \frac{\hat{s}}{\|\hat{s}\| + \varepsilon} \qquad (8.33)$$

where $\varepsilon > 0$ is small constant. This modification creates a small boundary layer around the switching surface in which the system trajectory will remain. Therefore, the chattering problem can be reduced significantly.

8.3. Simulation of a MEMS Gyroscope

We will evaluate the proposed adaptive sliding mode control with a sliding mode observer on the lumped MEMS gyroscope model. The control objective is to design an adaptive sliding mode controller so that the trajectory of $X(t)$ can track the state of reference model $X_m(t)$ and a consistent estimate of Ω_z can be obtained. In the simulation, we allowed $\pm$ 5% parameter variations for the spring and damping coefficients with respect to their nominal values and further assumed $\pm$ 2% magnitude changes in the coupling terms i.e. d_{xy} and ω_{xy}. Assuming the $\pm$ 5% time varying parameter variation of Ω_z. The external disturbance is a random variable with zero mean and unity variance. Parameters of the MEMS gyroscope are as follows:

$$m = 0.57e-8 \text{ kg}, \; d_{xx} = 0.429e-6 \text{ N s/m},$$

$$d_{xy} = 0.0429e-6 \text{ N s/m}, \; d_{yy} = 0.687e-6 \text{ N s/m},$$

$$k_{xx} = 80.98 \text{ N/m}, \; k_{xy} = 5 \text{ N/m}, \; k_{yy} = 71.62 \text{ N/m},$$

$$w_0 = 1kHz, \; q_0 = 10^{-6} m.$$

The unknown angular velocity is assumed $\Omega_z = 5.0$ rad/s and the initial condition on K matrix is $K(0) = 0.9K^*$, where $K^* = \begin{pmatrix} 14190 & 0.075 & 877.19 & -9.99 \\ 877.19 & 10 & 12539 & 12 \end{pmatrix}$. The desired motion trajectories are $x_m = \sin(w_1 t)$ and $y_m = 1.2\sin(w_2 t)$, where $w_1 = 4.17kHz$ and $w_2 = 5.11kHz$. The sliding mode matrix λ in (8.21) is chosen as

$\lambda = \begin{bmatrix} 0 & 10 & 0 & 0 \\ 0 & 0 & 0 & 10 \end{bmatrix}$, and matrix K_e is designed as $K_e = \begin{bmatrix} -10 & -10 & 10 & 10 \\ -10 & -10 & -10 & -10 \end{bmatrix}$. This last choice places the eigenvalues of the matrix $(A_m + BK_e)$ at $\{-8.10 \pm 11.44i \quad -1.90 \pm 1.40i\}$ (unit is rad/s) which corresponds the approximate damping ratios of 0.58 and 0.80 respectively. The sliding mode gain of (8.12) is $\rho_1 = 10000$. The adaptive gain of (8.31) is $M = diag\{20 \quad 20\}$. The observer feedback gain matrix L of (8.18) are designed as $L = \begin{bmatrix} 100 & 215 & 195 & 340 \\ 80 & 100 & 210 & 120 \end{bmatrix}^T$. The initial condition of the sliding mode observer in (8.18) is $\hat{X}(0) = [0.2 \quad 0.2 \quad 0.2 \quad 0.2]^T$ and the sliding gain of (8.15) is $\rho_2 = 4$. In the simulation of smooth sliding mode component of (8.33), $\varepsilon = 0.15$.

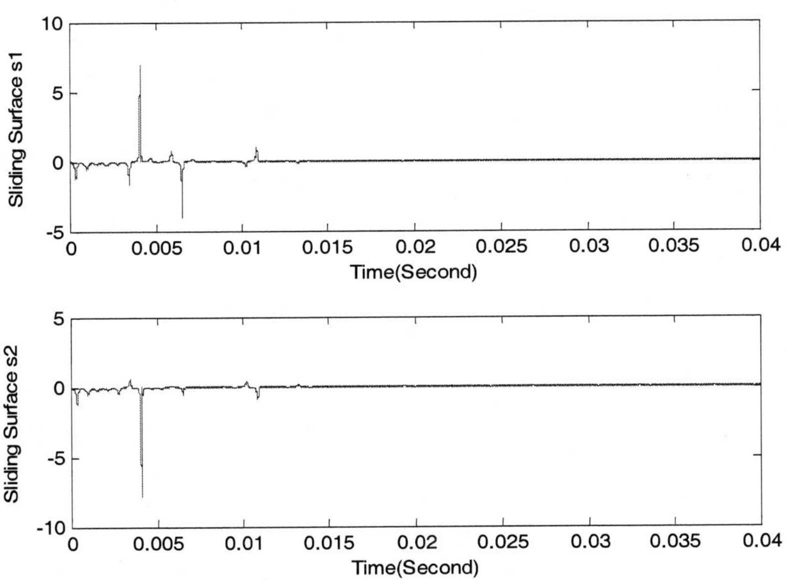

Figure 8.2. Convergence behavior of sliding surface $\hat{s}(t)$.

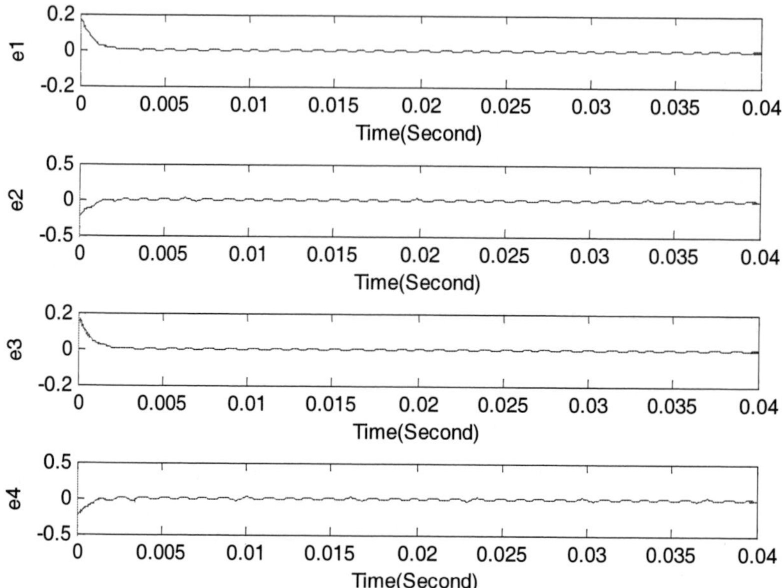

Figure 8.3. Convergence of the tracking error $\hat{e}(t)$.

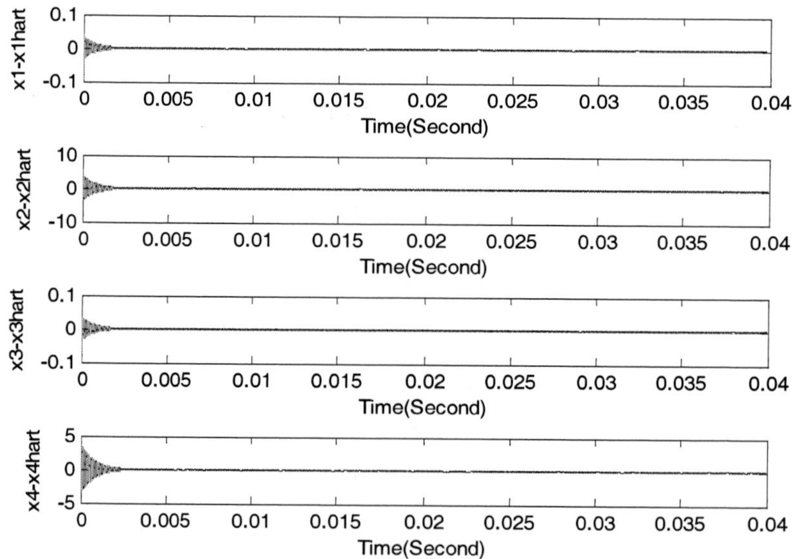

Figure 8.4. Convergence of the observation error $\widetilde{X}(t)$.

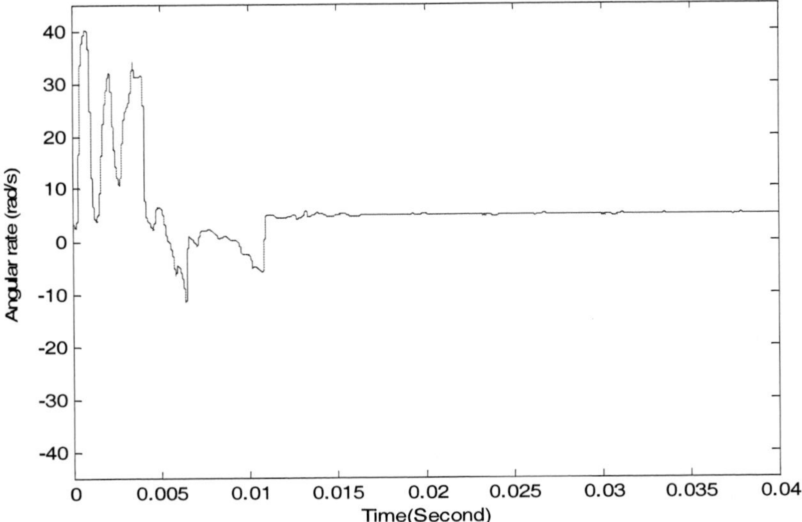

Figure 8.5. Convergence of the estimated angular rate Ω_z with the smooth sliding mode controller.

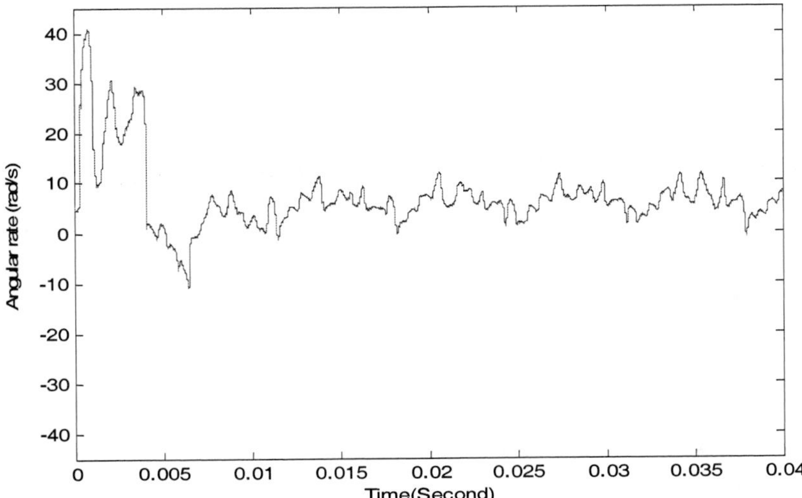

Figure 8.6. Convergence of the estimated angular rate Ω_z with the non-smooth sliding mode controller and observer.

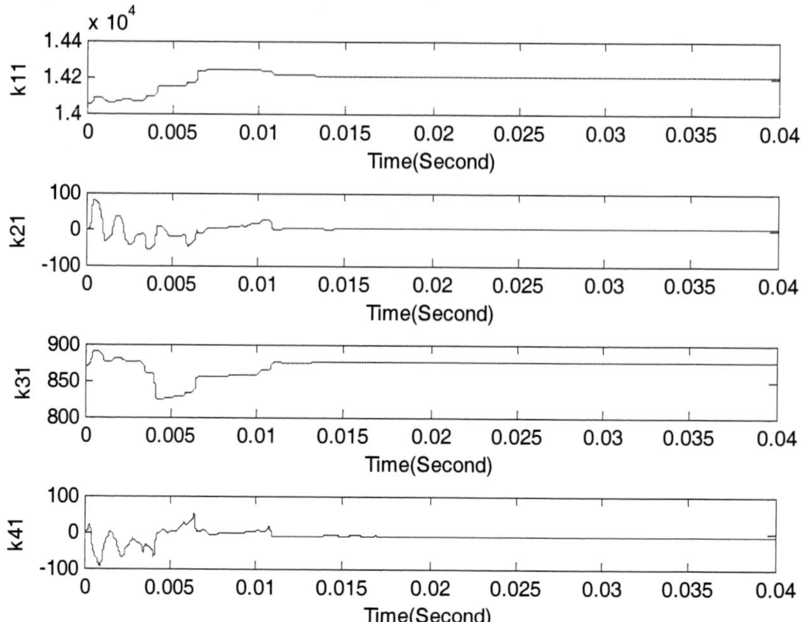

Figure 8.7. (Continued).

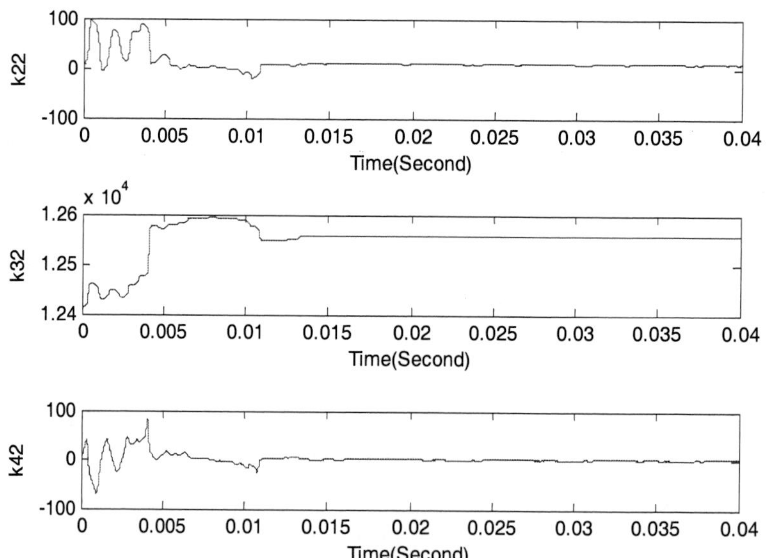

Figure 8.7. Estimations of controller parameters $K(t)$.

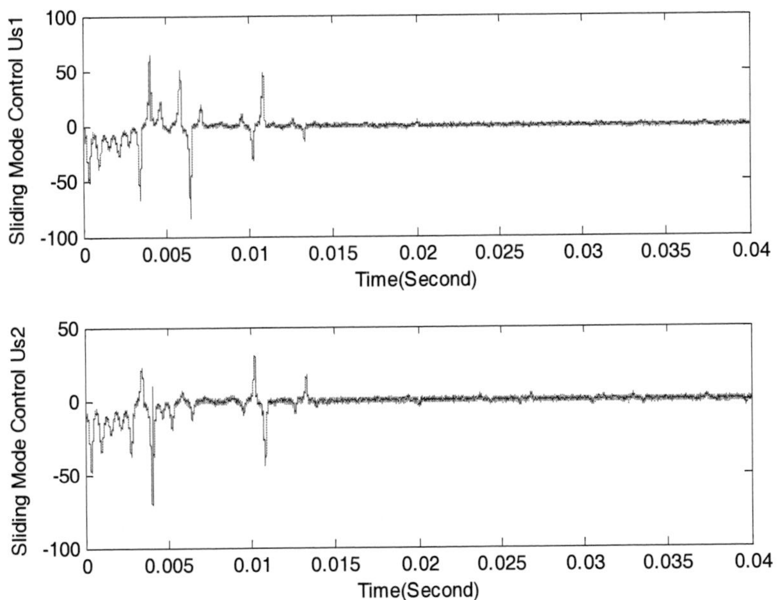

Figure 8.8. Smooth sliding mode control forces u_{s1} and u_{s2} of the adaptive sliding mode controller with the sliding mode observer.

CONCLUDING REMARKS

This chapter proposed an adaptive sliding mode controller with a proportional and integral sliding surface for a MEMS gyroscope. A nonlinear sliding mode observer that uses only the position signal was incorporated into the adaptive sliding mode control algorithm. All unmeasured states, as well as the angular velocity and unknown gyroscope parameters, are consistently estimated. The combined observer-controller synthesis involved three steps: first, a sliding mode controller was developed, assuming the availability of the state vector; second, a sliding mode observer of the state vector was designed to estimate the unmeasured states; and third, the sliding mode observer was combined with the proposed sliding mode controller, utilizing the estimate instead of the true state vector. Simulations demonstrated the robustness of the proposed adaptive sliding mode controller with the sliding mode observer. It was shown that the angular velocity and gyroscope parameters could be consistently estimated in the presence of model uncertainty, external disturbance and unmeasured states.

Chapter 9

COMPARATIVE STUDY OF ADAPTIVE VIBRATION CONTROL FOR MEMS GYROSCOPES

This chapter presents a comparative study of adaptive control approaches for MEMS z-axis gyroscopes. Novel adaptive controller and adaptive sliding mode controller are proposed, respectively, and the comparative analysis of these two methodologies is implemented. The difference in the controller derivation and stability analysis is discussed in detail. The proposed adaptive control approaches can estimate the angular velocity and the damping and stiffness coefficients including the coupling terms. The stability of the closed-loop systems are established with the proposed adaptive control strategies. Numerical examples are investigated to verify the effectiveness of the proposed control schemes. As compared with adaptive control, it is shown that adaptive sliding mode control could improve the robustness of the control system in the presence of an external disturbance.

In this chapter, a novel adaptive control is derived for the state tracking control of the MEMS gyroscope; moreover, for the purpose of comparison, a novel adaptive sliding mode control with an integral switching surface is designed to estimate the unknown parameters in the presence of external disturbance. The main advantage of the integral sliding surface is that it can provide more design flexibility and simplify the design procedure, meaning that the system dynamics in the sliding surface is only determined by reference model dynamics itself. The novelty of the proposed adaptive control lies in that an additional controller is incorporated into the state feedback controller, giving it more freedom to design the adaptive controller. Thus, the error dynamics is determined by the reference model dynamics and additional controller. In addition, a sliding mode control algorithm is incorporated into the proposed adaptive control and the adaptive sliding mode control, as applied to a MEMS gyroscope in the presence of external disturbance is investigated. Using Lyapunov stability theory and Barbalat's lemma, the stability of the closed-loop system and convergence properties can be guaranteed.

This chapter investigates the adaptive control approaches to estimate the angular velocity of MEMS gyroscope by using state tracking controllers. This chapter's primary contribution describes how novel adaptive approaches are proposed to estimate the angular velocity and all unknown gyroscope parameters. A comparative study of adaptive control and adaptive sliding mode control for the MEMS z-axis gyroscope is implemented.

9.1. PROBLEM FORMULATION

The gyroscope model in state space form can be represented as:

$$\dot{\boldsymbol{X}}(t) = \boldsymbol{A}\boldsymbol{X}(t) + \boldsymbol{B}\boldsymbol{u}(t) \tag{9.1}$$

where:

$$\boldsymbol{A} = \begin{bmatrix} 0 & 1 & 0 & 0 \\ -\omega_x^{\ 2} & -d_{xx} & -\omega_{xy} & -(d_{xy} - 2\Omega_z) \\ 0 & 0 & 0 & 1 \\ -\omega_{xy} & -(d_{xy} + 2\Omega_z) & -\omega_y^{\ 2} & -d_{yy} \end{bmatrix}$$

$$\boldsymbol{B} = \begin{bmatrix} 0 & 1 & 0 & 0 \\ 0 & 0 & 0 & 1 \end{bmatrix}^T, \quad \boldsymbol{u} = \begin{bmatrix} u_x \\ u_y \end{bmatrix}, \quad \boldsymbol{X} = \begin{bmatrix} x \\ \dot{x} \\ y \\ \dot{y} \end{bmatrix}.$$

The reference model $x_m = A_1\, sin(\omega_1 t)$, $y_m = A_2\, sin(\omega_2 t)$ is defined as:

$$\ddot{\boldsymbol{q}}_m + \boldsymbol{K}_m \boldsymbol{q}_m = 0 \tag{9.2}$$

where $\boldsymbol{K}_m = diag\{\omega_1^2 \quad \omega_2^2\}$.

The reference model can be written in state space form as:

$$\dot{\boldsymbol{X}}_m = \begin{bmatrix} 0 & 1 & 0 & 0 \\ -\omega_1^2 & 0 & 0 & 0 \\ 0 & 0 & 0 & 1 \\ 0 & 0 & -\omega_2^2 & 0 \end{bmatrix} \boldsymbol{X}_m \equiv \boldsymbol{A}_m \boldsymbol{X}_m \tag{9.3}$$

where $\boldsymbol{A}_m$ is a known constant matrix.

Consider the system in (9.1) with external disturbance as:

$$\dot{\boldsymbol{X}}(t) = \boldsymbol{AX}(t) + \boldsymbol{Bu}(t) + \boldsymbol{f}(t) \tag{9.4}$$

where $\boldsymbol{X}(t) \in R^4$, $\boldsymbol{u}(t) \in R^2$ and $\boldsymbol{A} \in R^{4\times4}$ is the unknown constant matrix, $\boldsymbol{B} \in R^{4\times2}$ is the known constant matrix, $\boldsymbol{f}(t)$ is an uncertain extraneous disturbance .

We make the following assumptions:

A1. The term $\boldsymbol{f}(t)$ has matched and unmatched terms. There exists an unknown matrix of an appropriate dimension $\boldsymbol{f}_m(t)$, such that:

$$\boldsymbol{f}(t) = \boldsymbol{Bf}_m(t) + \boldsymbol{f}_u(t) \tag{9.5}$$

where $\boldsymbol{Bf}_m(t)$ is a matched disturbance and $\boldsymbol{f}_u(t)$ is an unmatched disturbance.

Therefore, the dynamics (9.4) can be rewritten as:

$$\dot{\boldsymbol{X}}(t) = \boldsymbol{AX}(t) + \boldsymbol{Bu}(t) + \boldsymbol{Bf}_m + \boldsymbol{f}_u \tag{9.6}$$

A2. The matched and unmatched lumped uncertainty and external disturbance $\boldsymbol{f}_m$ and $\boldsymbol{f}_u$ are bounded, such as:

$$\|f_m(t)\| \le \alpha_m \text{ and } \|f_u(t)\| \le \alpha_u, \tag{9.7}$$

where α_m and α_u are known positive constants .

A3. There exists a constant matrix K^*, such that the matching condition, $A + BK^{*T} = A_m$, can always be satisfied.

Assumptions 1 and 2 are only effective for the adaptive sliding mode design in section 9.3 when in the presence of external disturbance, rather than the adaptive control design in section 9.2.

The control target for the MEMS gyroscope is (i) to design an adaptive controller so that the trajectory of $X(t)$ can track the state of reference model $X_m(t)$; (ii) to estimate the angular velocity of the MEMS gyroscope and all unknown gyroscope parameters.

9.2. Adaptive Control Design

In this section, an adaptive controller is proposed to estimate the angular velocity and all unknown gyroscope parameters. In the adaptive control design, we consider the equation (9.1) as the system model and use assumption 3.

The tracking error and its derivative are:

$$e(t) = X(t) - X_m(t) \tag{9.8}$$

$$\dot{e}(t) = A_m e(t) + (A - A_m)X(t) + Bu(t). \tag{9.9}$$

The adaptive controller is proposed as:

$$u(t) = K^T(t)X(t) + K_f e(t) \tag{9.10}$$

where $K(t)$ is an estimate of K^*, and the constant matrix K_f satisfies the condition that $(A_m + BK_f)$ is Hurwitz.

Defining the estimation error as $\tilde{K}(t) = K(t) - K^*$ and then substituting (9.10) into (9.1) yields:

$$\dot{X}(t) = A_m X(t) + B\tilde{K}^T(t)X(t) \ . \tag{9.11}$$

Then, we have the tracking error equation:

$$\dot{e}(t) = (A_m + BK_f)e(t) + B\tilde{K}^T(t)X(t) \ . \tag{9.12}$$

The updated law is derived based on the state $X(t)$, and the tracking error $e(t)$, shown as follows:

$$\dot{\tilde{K}}^T(t) = \dot{K}^T(t) = -MB^T P^T e(t)X^T(t) \tag{9.13}$$

where $M = diag\{m_1 \quad m_2\}$ is positive definite matrix.

The stability analysis of the proposed adaptive controller is summarized in Theorem 1:

Theorem 1. The adaptive controller (9.10) with the adaptive law (9.13), applied to the system (9.1), guarantees that all closed-loop signals are bounded, the tracking error goes to zero asymptotically, and the controller parameter K converges to its true value if the condition of persistent excitation is satisfied.

Proof: Define a Lyapunov function:

$$V = \frac{1}{2}e^T Pe + \frac{1}{2}tr\left[\tilde{K}M^{-1}\tilde{K}^T\right] \tag{9.14}$$

where P is positive definite matrix.

Differentiating V, with respect to time, yields:

$$\begin{aligned} \dot{V} &= e^T P\dot{e} + tr\left[\tilde{K}M^{-1}\dot{\tilde{K}}^T\right] \\ &= -e^T Qe + e^T PB\tilde{K}^T X + tr\left[\tilde{K}M^{-1}\dot{\tilde{K}}^T\right] \end{aligned} \tag{9.15}$$

where $P(A_m + BK_f) + (A_m + BK_f)^T P = -Q$, Q is positive definite matrix.

Choose the adaptive law as (21) with $K(0)$ being arbitrary yields:

$$\dot{V} = -e^T Qe \leq -\lambda_{min}(Q)\|e\| \leq 0 . \tag{9.16}$$

The inequality $\dot{V} \leq -\lambda_{min}(Q)\|e\|$ implies that e is integrable as $\int_0^t \|e\|dt \leq \frac{1}{\lambda_{min}(Q)}[V(0)-V(t)]$. Since $V(0)$ is bounded and $V(t)$ is nonincreasing and bounded, it can be concluded that $\lim_{t\to\infty}\int_0^t \|e\|dt$ is bounded. Since $\lim_{t\to\infty}\int_0^t \|e\|dt$ is bounded and $\dot{e}$ is also bounded, according to Barbalat's lemma, e will asymptotically converge to zero, $\lim_{t\to\infty} e(t) = 0$.

In order to show persistence of excitation, we need to show that there exists some positive scalar constants α and T, such that for all $t>0$ $\int_t^{t+T} XX^T d\tau \geq \alpha I$. It can be shown that there always exists some positive scalar constants α and T, such that for all $t>0$, $\int_t^{t+T} XX^T d\tau \geq \alpha I$ if XX^T is non-singular matrix. If $X(t)$ is a persistent excitation signal, it can be guaranteed that $\tilde{K}(t) \to 0$.

Remark 1. If persistently exciting drive signals, $x_m = A_1 sin(\omega_1 t)$ and $y_m = A_2 sin(\omega_2 t)$ are used, where $w_1 \neq w_2$, then $\tilde{K}(t)$ and $e(t)$ all converge to zero asymptotically. Consequently, the unknown angular velocity as well as other unknown parameters can be determined from $A + BK^T = A_m$ and the unknown angular velocity can be determined by $\Omega_z = 0.25(k_{22} - k_{41})$ as $lim_{t\to\infty} \Omega_z(t) = \Omega_z$. It is difficult to establish the convergence rate, however.

9.3. Adaptive Sliding Mode Control

A novel adaptive sliding mode control strategy, with a proportional and integral sliding surface for MEMS gyroscopes, is proposed in this section. A detailed study of the proportional-integral sliding mode control algorithm is presented in the presence of both matched and mismatched external disturbances.

In the adaptive sliding mode control design, if we consider equation (9.4) and define the tracking error $e(t) = X(t) - X_m(t)$, then its derivative becomes:

$$\dot{e}(t) = A_m e(t) + (A - A_m)X(t) + Bu(t) + Bf_m(t) + f_u(t). \quad (9.18)$$

The proportional-integral sliding surface $s = 0$ is defined as:

$$s(t) = \lambda e(t) - \int_0^t \lambda(A_m + BK_e)e(t)d\tau \quad (9.19)$$

where λ is a constant matrix which satisfies that λB is nonsingular. The constant K_e satisfies the condition that $(A_m + BK_e)$ is Hurwitz.

The derivative of the sliding surface is:

$$\dot{s}(t) = \lambda(A - A_m)X(t) + \lambda Bu(t) + \lambda Bf_m(t) + \lambda f_u(t) - \lambda BK_e e(t) \quad (9.20)$$

Setting $\dot{s} = 0$ to solve equivalent control u_{eq} gives:

$$\begin{aligned} u_{eq}(t) &= -(\lambda B)^{-1}\lambda(A - A_m)X(t) + K_e e(t) - f_m(t) - (\lambda B)^{-1}\lambda f_u(t) \\ &= K^{*T}X(t) + K_e e(t) - f_m(t) - (\lambda B)^{-1}\lambda f_u(t). \end{aligned} \quad (9.21)$$

The adaptive sliding mode controller is proposed as:

$$u(t) = K^T(t)X(t) + K_e e(t) - \rho(\lambda B)^{-1}\frac{s(t)}{\|s(t)\|} \quad (9.22)$$

where $K(t)$ is and estimate of K^*. The last component of the control signal $u(t)$ is designed to address the matched and unmatched disturbance. This component is given as $u_s = \begin{bmatrix} u_{s1} \\ u_{s2} \end{bmatrix} = -\rho(\lambda B)^{-1}\frac{s}{\|s\|}$, where ρ is a constant.

Define the estimation error as:

$$\widetilde{K}(t) = K(t) - K^* . \tag{9.23}$$

Substituting (9.22) and (9.23) into (9.6) yields:

$$\dot{X}(t) = A_m X(t) + B\widetilde{K}^T(t)X(t) + BK_e e(t) + Bf_m(t) + f_u(t) - B\rho(\lambda B)^{-1}\frac{s(t)}{\|s(t)\|}. \tag{9.24}$$

Then, we have the tracking error equation:

$$\dot{e}(t) = (A_m + BK_e)e(t) + B\widetilde{K}^T(t)X(t) + Bf_m(t) + f_u(t) - B\rho(\lambda B)^{-1}\frac{s(t)}{\|s(t)\|}. \tag{9.25}$$

Therefore, the derivative of $s(t)$ becomes:

$$\dot{s}(t) = \lambda B\widetilde{K}^T(t)X(t) + \lambda Bf_m(t) + \lambda f_u(t) - \rho\frac{s(t)}{\|s(t)\|}. \tag{9.26}$$

The updated law for the estimated parameters is derived, based on the state $X(t)$, and sliding surface $s(t)$, as follows:

$$\dot{\widetilde{K}}^T(t) = \dot{K}^T(t) = -MB^T\lambda^T sX^T(t) \tag{9.27}$$

The stability analysis of the proposed adaptive sliding mode control can be summarized in Theorem 2:

Theorem 2. The adaptive controller (9.22) with the adaptive law (9.29), applied to the system (9.4) guarantees that all closed-loop signals are bounded, the tracking error and sliding surface go to zero asymptotically, and the controller parameter K converges to its true value if the condition of persistent excitation is satisfied.

Proof: Define a Lyapunov function:

$$V = \frac{1}{2} s^T s + \frac{1}{2} tr\left[\tilde{K} M^{-1} \tilde{K}^T\right]. \tag{9.28}$$

Differentiating V, with respect to time, yields:

$$\begin{aligned}
\dot{V} &= s^T \dot{s} + tr\left[\tilde{K} M^{-1} \dot{\tilde{K}}^T\right] \\
&= s^T \left[\lambda B \tilde{K}^T(t) X(t) + \lambda B f_m + \lambda f_u - \rho \frac{s}{\|s\|} \right] + tr\left[\tilde{K} M^{-1} \dot{\tilde{K}}^T\right] \\
&= -s^T \rho \frac{s}{\|s\|} + s^T \lambda B f_m + s^T \lambda f_u + s^T \lambda B \tilde{K}^T(t) X(t) + tr\left[\tilde{K} M^{-1} \dot{\tilde{K}}^T\right]
\end{aligned} \tag{9.29}$$

Choose the adaptive law as (9.29) with $K(0)$ being arbitrary yields:

$$\begin{aligned}
\dot{V} &= -\rho\|s\| + s^T \lambda B f_m + s^T \lambda f_u \\
&\le -\rho\|s\| + \|s\|\|\lambda B\|\|f_m\| + \|s\|\|\lambda\|\|f_u\| \\
&\le -\rho\|s\| + \|s\|\|\lambda B\|\alpha_m + \|s\|\|\lambda\|\alpha_u \\
&= -\|s\|\left[\rho - \|\lambda B\|\alpha_m - \|\lambda\|\alpha_u\right]
\end{aligned} \tag{9.30}$$

with the choice of $\rho \ge \|\lambda B\|\alpha_m + \|\lambda\|\alpha_u + \eta$ where η is a positive constant, and $\dot{V}$ becomes negative semi-definite, i.e., $\dot{V} \le -\eta\|s\|$. This implies that the trajectory reaches the sliding surface in finite time and remains on the sliding surface. $\dot{V}$ is negative definite, implying that s and $\tilde{K}$ converge to zero. $\dot{V}$ is negative semi-definite, ensuring that V, s and $\tilde{K}$ are all bounded. It can be concluded from (9.29) that $\dot{s}$ is also bounded. The inequality $\dot{V} \le -\eta\|s\|$ implies that s is integrable as $\int_0^t \|s\| dt \le \frac{1}{\eta}\left[V(0) - V(t)\right]$. Since $V(0)$ is bounded and $V(t)$ is nonincreasing and bounded, it can be concluded that $\lim_{t\to\infty} \int_0^t \|s\| dt$ is bounded. Since $\lim_{t\to\infty} \int_0^t \|s\| dt$ is bounded and $\dot{s}$ is also bounded, according to Barbalat lemma, $s(t)$ will asymptotically converge to zero, $\lim_{t\to\infty} s(t) = 0$. Then $e(t)$ will also asymptotically converge to zero.

If X is a persistent excitation signal, then $\dot{\tilde{K}}^T(t) = -MB^T\lambda^T sX^T$ guarantees $\tilde{K} \to 0$, K will converge to its true value. It can be shown that if $\omega_1 \neq \omega_2$, there always exists some positive scalar constants α and T, such that for all $t > 0$, $\int_t^{t+T} XX^T d\tau \geq \alpha I$, where

$$XX^T = \begin{bmatrix} x_1^2 & x_1\dot{x}_1 & x_1x_2 & x_1\dot{x}_2 \\ \dot{x}_1x_1 & \dot{x}_1^2 & \dot{x}_1x_2 & \dot{x}_1\dot{x}_2 \\ x_2x_1 & x_2\dot{x}_1 & x_2^2 & x_2\dot{x}_2 \\ \dot{x}_2x_1 & \dot{x}_2\dot{x}_1 & \dot{x}_2x_2 & \dot{x}_2^2 \end{bmatrix}.$$

It can be shown that XX^T has full rank if $\omega_1 \neq \omega_2$, i.e. the excitation frequencies on x and y axes should be different. Therefore, it can be concluded that $\tilde{K}(t)$, $s(t)$, and $e(t)$ all converge to zero asymptotically.

Remark 1. In the proposed adaptive control and proposed adaptive sliding mode control, both of the tracking errors are state tracking errors, but their derivatives are different because only the external disturbances are considered in the adaptive sliding mode control design, the equation for the error dynamics is completely different between these two control methodologies.

9.4. Simulation Example

We will evaluate the proposed adaptive control and adaptive sliding mode control on the lumped MEMS gyroscope model. The control objective is to design an adaptive state tracking controller so that a consistent estimate of angular velocity can be obtained.

In the simulation, with respect to their nominal values, we allowed $\pm$ 10% parameter variations for the spring and damping coefficients. We further assumed $\pm$ 10% magnitude changes in the coupling terms, i.e. d_{xy} and ω_{xy}, again with respect to their nominal values. The external disturbance is a random variable signal with a zero mean and unity variance. Parameters of the MEMS gyroscope are as follows:

$m = 0.57e^{-8}$ kg, $d_{xx} = 0.429e^{-6}$ N s/m, $d_{xy} = 0.0429e^{-6}$ N s/m, $d_{yy} = 0.687e^{-6}$ N s/m,

$k_{xx} = 80.98$ N/m, $k_{xy} = 5$ N/m, $k_{yy} = 71.62$ N/m, $w_0 = 1kHz$, $q_0 = 10^{-6} m$.

The unknown angular velocity is assumed to be $\Omega_z = 5.0$ rad/s and the initial condition on $\boldsymbol{K}$ matrix is $\boldsymbol{K}(0) = 0.95\boldsymbol{K}^*$. The desired motion trajectories are $x_m = \sin(w_1 t)$ and $y_m = 1.2\sin(w_2 t)$, where $w_1 = 6.71kHz$ and $w_2 = 5.11kHz$. The adaptive gain is $\boldsymbol{M} = diag\{20 \quad 20\}$. $\boldsymbol{K}_f = \begin{bmatrix} -10000 & -10000 & 1000 & 20000 \\ -1000 & -1000 & -1000 & -1000 \end{bmatrix}$, $\boldsymbol{K}_e = \begin{bmatrix} -10000 & -10000 & 1000 & 20000 \\ -1000 & -1000 & -1000 & -1000 \end{bmatrix}$, $\boldsymbol{\lambda} = \begin{bmatrix} 0 & 10 & 0 & 0 \\ 0 & 0 & 0 & 10 \end{bmatrix}$.

The sliding mode gain is chosen as $\rho = diag\{10000 \quad 10000\}$ to satisfy the stability condition. The smooth sliding mode component is $\varepsilon = 0.01$.

Figures 9.1 and 9.2 compare the tracking errors. It can be observed that the tracking errors all converge to zero asymptotically and the tracking error of adaptive sliding mode control is smaller than that of adaptive control. Figures 9.3 to 9.6 compare the adaptation of the angular velocity and controller parameters using these two different controllers. It is graphed that the estimates of angular velocity and controller parameters, by using an adaptive sliding mode control, have better performance, as compared with the adaptive control.

It can be observed from Figures 9.7 and 9.8 that the adaptive sliding mode control input is larger than that of the adaptive control input. Figure 9.9 depicts the sliding surface and demonstrates that the sliding surfaces converge to zero asymptotically. It can be noticed in Figure 9.10 that the chattering problem has been reduced by using the smooth sliding mode controller.

The estimate of angular velocity by using an adaptive sliding mode control has a larger overshoot at the beginning, but a much smaller rise time than by using adaptive control. The external disturbance is difficult to compensate for in the adaptive controller; whereas the disturbance term can be considered in the adaptive sliding mode control. Therefore, when in the presence of external disturbance, the adaptive sliding mode control is better than adaptive control.

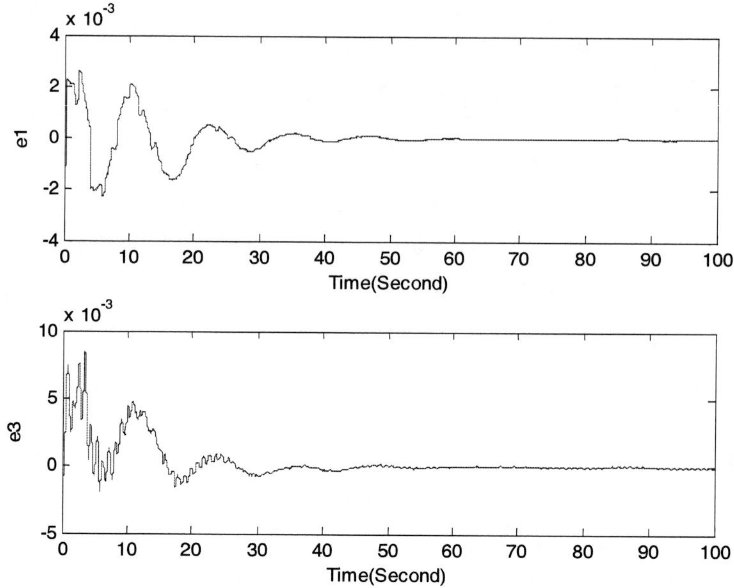

Figure 9.1. The tracking error using adaptive control.

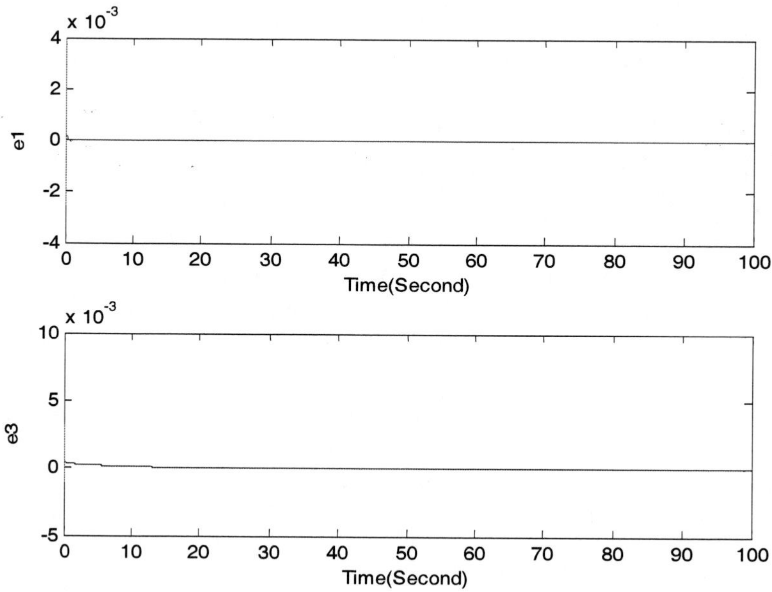

Figure 9.2. The tracking error using adaptive sliding mode control.

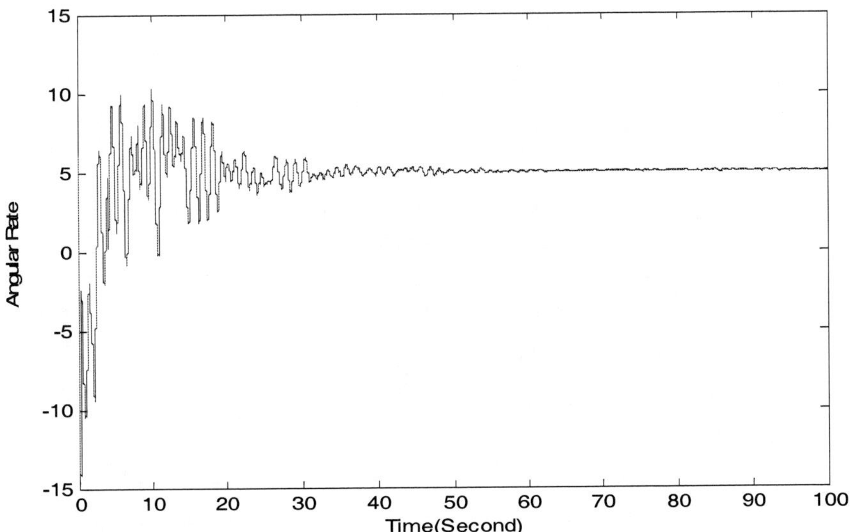

Figure 9.3. Adaptation of angular velocity using adaptive control.

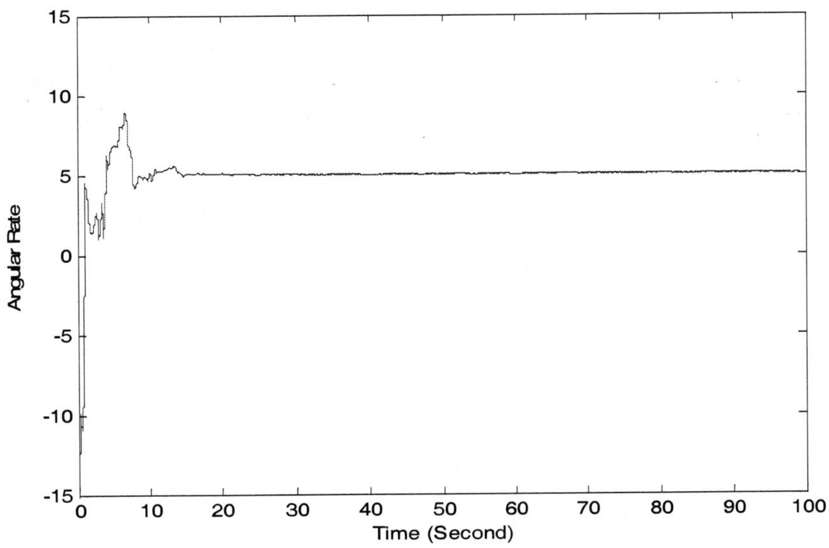

Figure 9.4. Adaptation of angular velocity using adaptive sliding mode control.

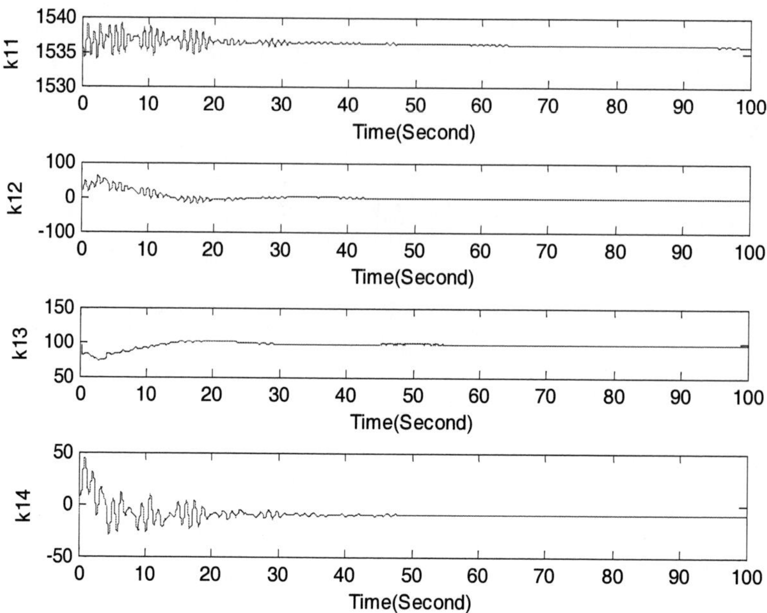

Figure 9.5. (Continued).

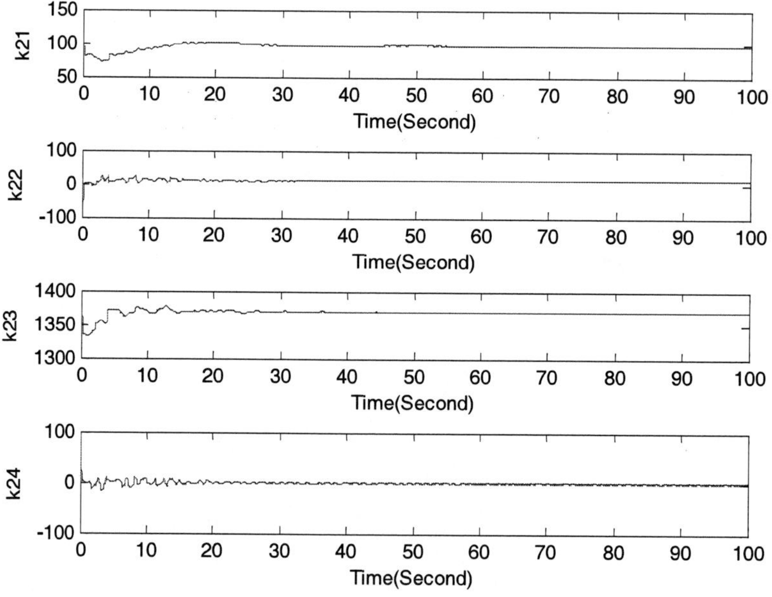

Figure 9.5. Adaptation of control parameters using adaptive control.

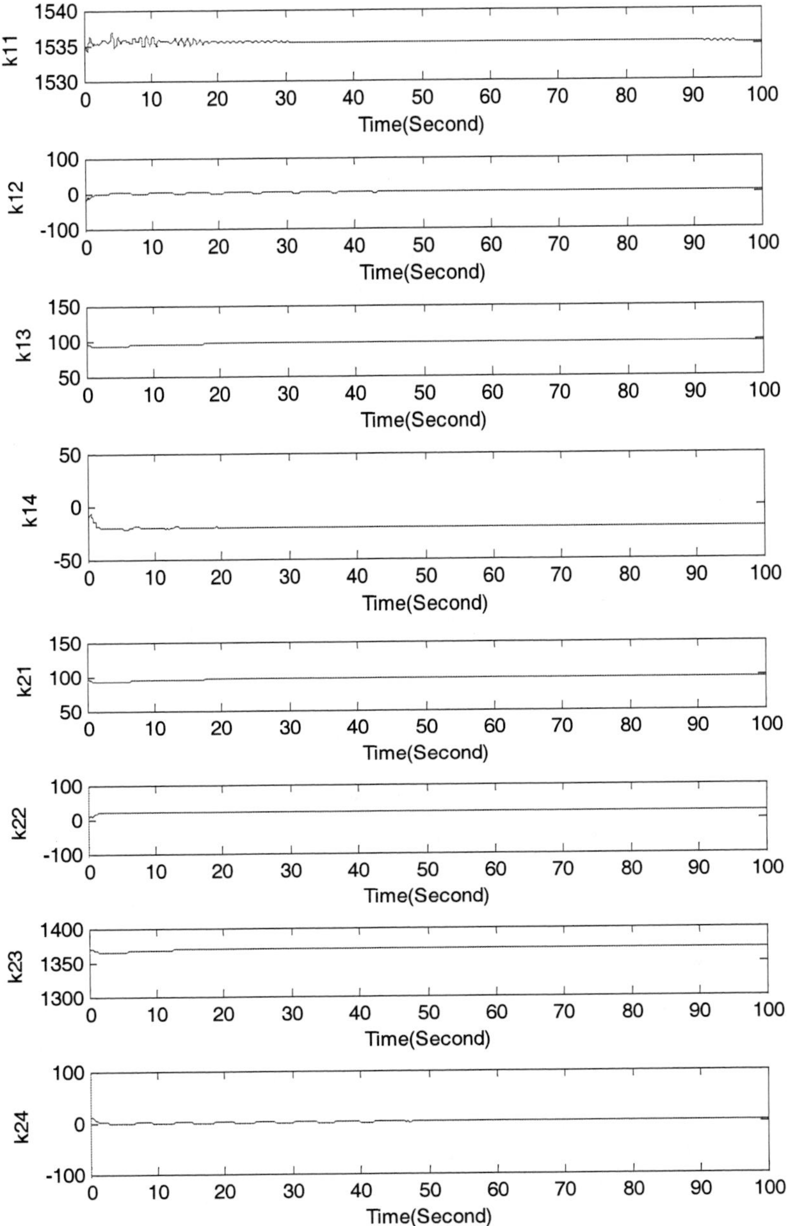

Figure 9.6. Adaptation of control parameters using adaptive sliding mode control.

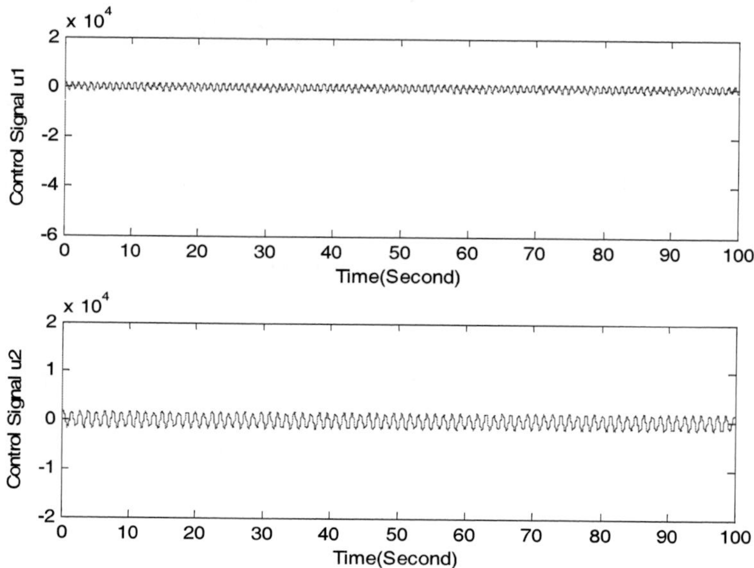

Figure 9.7. Adaptive control input.

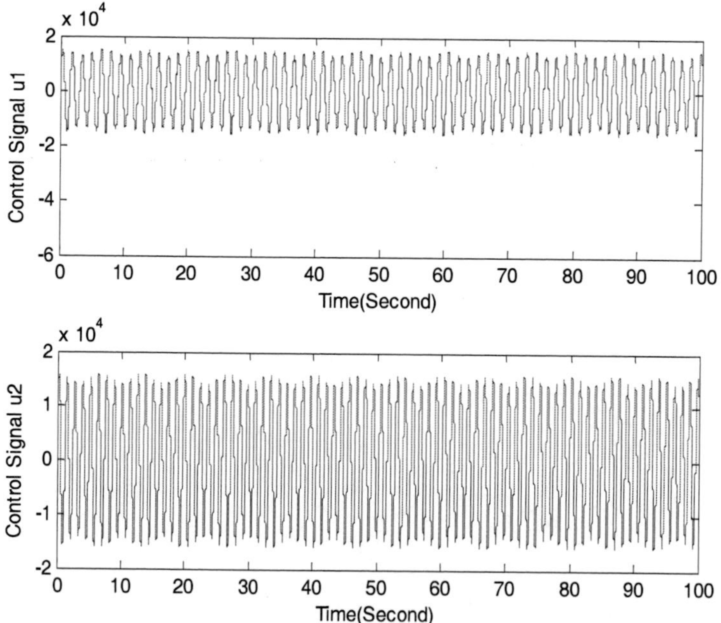

Figure 9.8. Adaptive sliding mode control input.

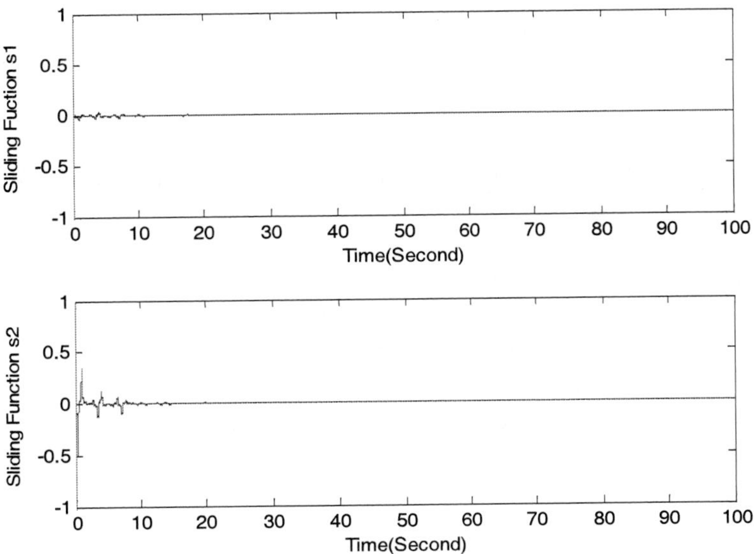

Figure 9.9. Convergence of the sliding surface **s**(t).

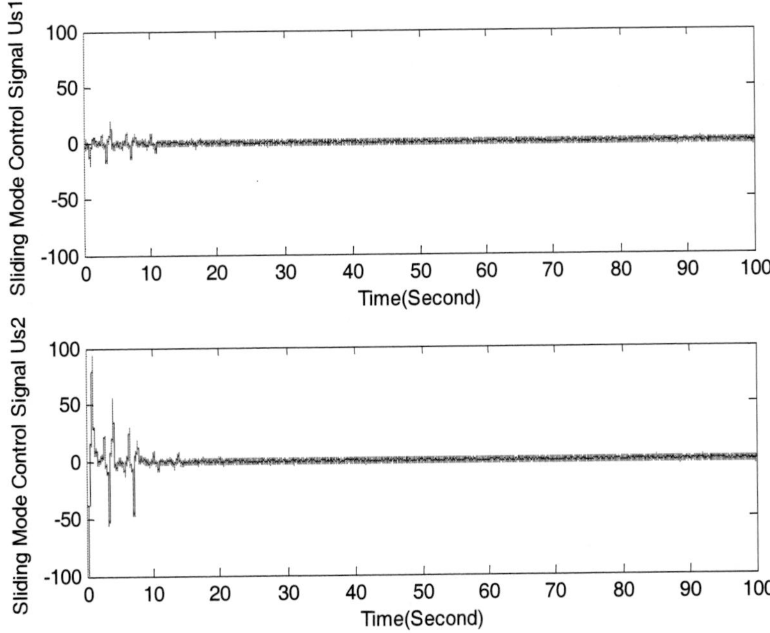

Figure 9.10. The smooth sliding mode control input.

CONCLUDING REMARKS

This chapter investigated the designs of adaptive control and adaptive sliding mode control for MEMS gyroscopes. Novel adaptive approaches were proposed and Lyapunov stability conditions were established. The difference between these two adaptive approaches is that the external disturbance is considered in the derivation of adaptive sliding mode algorithm and the robustness can be improved. Simulation results demonstrated that the effectiveness of the proposed adaptive control techniques when identifying the gyroscope parameters and angular velocity. Moreover, as compared to adaptive control, it could be concluded that adaptive sliding mode control has better robustness in the presence of external disturbance.

Chapter 10

ROBUST ADAPTIVE SLIDING MODE CONTROL OF MEMS TRIAXIAL GYROSCOPE

This chapter presents a robust tracking control strategy by using an adaptive sliding mode approach for a MEMS triaxial angular sensor device, able to detect rotation in three orthogonal axes and using a single vibrating mass. An adaptive sliding mode controller with proportional and integral sliding surface is developed and the stability of the closed-loop system can be guaranteed with the proposed adaptive sliding mode control strategy. The proposed adaptive sliding mode controller updates estimates of all stiffness errors, damping and input rotation parameters in real time, thereby removing the need for any offline calibration stages. The necessary model trajectory, to enable all unknown parameter estimates and to converge to their true values, is shown to be a three-dimensional Lissajous pattern..

This chapter extends the adaptive sliding mode control from two axial angular sensors to triaxial angular sensors, proposing a novel concept for an adaptively controlled triaxial angular velocity sensor device that is able to detect rotation in three orthogonal axes and using a single vibrating mass. The use of a single mass to sense triaxial rotation promises to avoid mechanical interference, as well as reduce costs and energy consumption. The triaxial angular velocity sensor will be based on a surface micromachining technology, capable of sensing the angular motion of about three orthogonal axes. It provides analog outputs for angular velocity and position references about the X, Y, and Z-axes.

10.1. Adaptive Sliding Mode Controller Design

This section proposes a new adaptive sliding mode control strategy for MEMS triaxial gyroscopes, using a proportional and integral sliding surface, as shown in Figure 10.1. A detailed study of the proportional-integral sliding mode control algorithm is presented in the presence of matched and mismatched uncertainties and external disturbance with the triaxial gyroscope model. The control target for a MEMS gyroscope is to maintain the proof mass to oscillate in the x 、 y and z directions at given frequency and amplitude $x_m = A_1 \sin(w_1 t)$, $y_m = A_2 \sin(w_2 t)$, $z_m = A_3 \sin(w_3 t)$.

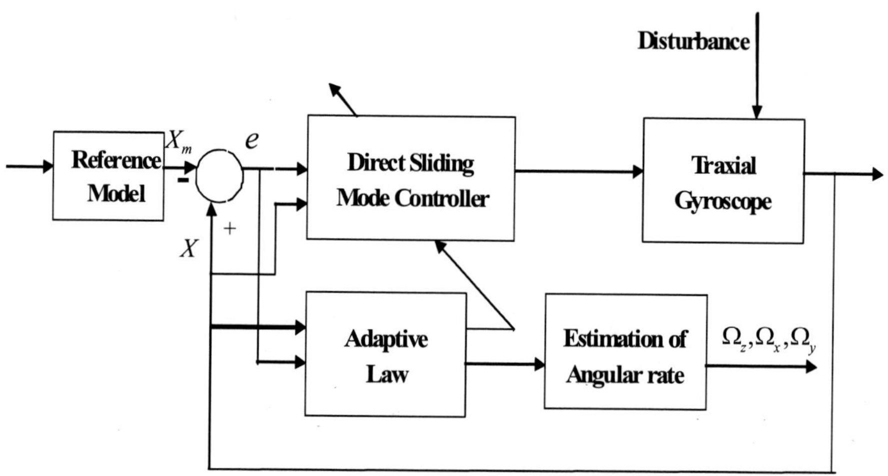

Figure 10.1. The block diagram of the adaptive sliding mode control for the triaxial MEMS gyroscope.

Rewriting the gyroscope model in state space equation:

$$\dot{X} = \begin{bmatrix} 0 & 1 & 0 & 0 & 0 & 0 \\ -w_x^2 & -d_{xx} & -w_{xy} & -(d_{xy} - 2\Omega_z) & -w_{xz} & -(d_{xz} + 2\Omega_y) \\ 0 & 0 & 0 & 1 & 0 & 0 \\ -w_{xy} & -(d_{xy} + 2\Omega_z) & -w_y^2 & -d_{yy} & -w_{yz} & -(d_{yz} - 2\Omega_x) \\ 0 & 0 & 0 & 0 & 0 & 1 \\ -w_{xz} & -(d_{xz} - 2\Omega_y) & -w_{yz} & -(d_{yz} + 2\Omega_x) & -w_z^2 & -d_{zz} \end{bmatrix} X + \begin{bmatrix} 0 & 0 & 0 \\ 1 & 0 & 0 \\ 0 & 0 & 0 \\ 0 & 1 & 0 \\ 0 & 0 & 0 \\ 0 & 0 & 1 \end{bmatrix} \begin{bmatrix} u_x \\ u_y \\ u_z \end{bmatrix} \tag{10.1}$$

which is $\dot{X} = AX + Bu$,where $X = \begin{bmatrix} x & \dot{x} & y & \dot{y} & z & \dot{z} \end{bmatrix}^T$.

The reference model is defined as:

$$\ddot{q}_m + K_m q_m = 0 \tag{10.2}$$

where $K_m = diag\{w_1^2 \quad w_2^2 \quad w_3^2\}$.

Rewriting the reference model in state space equation:

$$\dot{X}_m = \begin{bmatrix} 0 & 1 & 0 & 0 & 0 & 0 \\ -w_1^2 & 0 & 0 & 0 & 0 & 0 \\ 0 & 0 & 0 & 1 & 0 & 0 \\ 0 & 0 & -w_2^2 & 0 & 0 & 0 \\ 0 & 0 & 0 & 0 & 0 & 1 \\ 0 & 0 & 0 & 0 & -w_3^2 & 0 \end{bmatrix} X_m \tag{10.3}$$

which is $\dot{X}_m = A_m X_m$, where $X_m = [x_m \quad \dot{x}_m \quad y_m \quad \dot{y}_m \quad z_m \quad \dot{z}_m]^T$.

Consider the system with parametric uncertainties:

$$\dot{X}(t) = (A + \Delta A)X(t) + Bu + f(t) \tag{10.4}$$

where $X(t) \in R^n$, $u(t) \in R^m$ and $A \in R^{n \times n}$ is an unknown matrix, ΔA is the unknown parameter uncertainties of the matrix A, and $f(t)$ is an uncertain extraneous disturbance . We make the following assumptions:

A1. ΔA and $f(t)$ have matched and unmatched terms. There exist unknown matrices of appropriate dimensions D , G , such that $\Delta A(t) = BD(t) + \Delta\tilde{A}(t)$ and $f(t) = BG(t) + \tilde{f}(t)$, where $BD(t)$ is a matched uncertainty and $\Delta\tilde{A}(t)$ is an unmatched uncertainty, $BG(t)$ is a matched disturbance and $\tilde{f}(t)$ is an unmatched disturbance.

From this assumption, (10.4) can be rewritten as:

$$\begin{aligned} \dot{X}(t) &= AX(t) + Bu(t) + \Delta AX(t) + f(t) \\ &= AX(t) + Bu(t) + BDX(t) + \Delta\tilde{A}X(t) + BG + \tilde{f}(t) \\ &= AX(t) + Bu(t) + Bf_m + f_u \end{aligned} \tag{10.5}$$

where $f_m(t,X,u)$ represents the matched lumped uncertainty and disturbance, which is given by:

$$f_m(t,X)=DX(t)+G. \tag{10.6}$$

The term $f_u(t,X)$ represents the system lumped unmatched uncertainty and disturbance, which is given by:

$$f_u(t,X)=\Delta\tilde{A}X(t)+\tilde{f}(t). \tag{10.7}$$

A2. The matched and unmatched lumped uncertainty and external disturbance f_m and f_u are bounded, such as $\|f_m(t,X)\|\le\alpha_{m1}\|X\|+\alpha_{m2}$ and $\|f_u(t,X)\|\le\alpha_{u1}\|X\|+\alpha_{u2}$, where α_{m1}, α_{m2}, α_{u1}, α_{u2} are known positive constants .

A3. There exists a constant matrix K^*, such that the following matching condition $A+BK^{*T}=A_m$ can always be satisfied.

Remark 1. From $A+BK^{*T}=A_m$, we get $K^{*T}=(B^TB)^{-1}B^T(A_m-A)$,

$$A_m-A=\begin{bmatrix} 0 & 0 & 0 & 0 & 0 & 0 \\ -w_1^2+w_x^2 & d_{xx} & w_{xy} & d_{xy}-2\Omega_z & w_{xz} & d_{xy}+2\Omega_y \\ 0 & 0 & 0 & 0 & 0 & 0 \\ w_{xy} & d_{xy}+2\Omega_z & -w_2^2+w_y^2 & d_{yy} & w_{yz} & d_{yz}-2\Omega_x \\ 0 & 0 & 0 & 0 & 0 & 0 \\ w_{xz} & d_{xy}-2\Omega_y & w_{yz} & d_{yz}+2\Omega_x & -w_3^2+w_z^2 & d_{zz} \end{bmatrix}$$

Thus, from $K^{*T}=(B^TB)^{-1}B^T(A_m-A)$, i.e.:

$$\begin{bmatrix} k_{11}^* & k_{21}^* & k_{31}^* & k_{41}^* & k_{51}^* & k_{61}^* \\ k_{12}^* & k_{22}^* & k_{32}^* & k_{42}^* & k_{52}^* & k_{62}^* \\ k_{13}^* & k_{23}^* & k_{33}^* & k_{43}^* & k_{53}^* & k_{63}^* \end{bmatrix}$$

$$= \begin{bmatrix} -w_1^2 + w_x^2 & d_{xx} & w_{xy} & d_{xy} - 2\Omega_z & w_{xz} & d_{xy} + 2\Omega_y \\ w_{xy} & d_{xy} + 2\Omega_z & -w_2^2 + w_y^2 & d_{yy} & w_{yz} & d_{yz} - 2\Omega_x \\ w_{xz} & d_{xy} - 2\Omega_y & w_{yz} & d_{yz} + 2\Omega_x & -w_3^2 + w_z^2 & d_{zz} \end{bmatrix}$$

We can obtain the parameters of MEMS gyroscope as follows:

$d_{xx} = k_{21}$, $d_{yy} = k_{42}$, $d_{zz} = k_{63}$, $w_x^2 = k_{11} + w_1^2$, $w_y^2 = k_{32} + w_2^2$, $w_z^2 = k_{53} + w_3^2$,

$w_{xy} = k_{31} = k_{12}$, $w_{yz} = k_{52} = k_{33}$, $w_{xz} = k_{51} = k_{13}$,

$\Omega_x = 0.25(k_{43} - k_{62})$, $\Omega_y = 0.25(k_{61} - k_{23})$, $\Omega_z = 0.25(k_{22} - k_{41})$,

$d_{xy} = 0.5(k_{41} + k_{22})$, $d_{yz} = 0.5(k_{43} + k_{62})$, $d_{xz} = 0.5(k_{61} + k_{23})$.

The tracking error and its derivative are:

$$e(t) = X(t) - X_m(t) \tag{10.8}$$

$$\dot{e} = A_m e + (A - A_m)X + Bu + Bf_m + f_u \ . \tag{10.9}$$

The proportional-integral sliding surface is defined as:

$$s(t) = \lambda e - \int_0^t \lambda(A_m + BK_e)e d\tau \tag{10.10}$$

where $\lambda = \begin{bmatrix} \lambda_{11} & \lambda_{12} & \lambda_{13} & \lambda_{14} & \lambda_{15} & \lambda_{16} \\ \lambda_{21} & \lambda_{22} & \lambda_{23} & \lambda_{24} & \lambda_{25} & \lambda_{26} \\ \lambda_{31} & \lambda_{32} & \lambda_{33} & \lambda_{34} & \lambda_{35} & \lambda_{36} \end{bmatrix}$ is a constant matrix which satisfies that λB is a nonsingular diagonal matrix, and K_e is a constant matrix which satisfies that $(A_m + BK_e)$ is Hurwitz.

The derivative of the sliding surface is:

$$\dot{s} = \lambda(A - A_m)X + \lambda Bu + \lambda Bf_m + \lambda f_u - \lambda BK_e e. \tag{10.11}$$

Setting $\dot{s} = 0$ to solve equivalent control u_{eq} gives:

$$\begin{aligned} u_{eq} &= -(\lambda B)^{-1}\lambda(A - A_m)X + K_e e - f_m - (\lambda B)^{-1}\lambda f_u \\ &= K^{*T}X(t) + K_e e - f_m - (\lambda B)^{-1}\lambda f_u. \end{aligned} \tag{10.12}$$

From (10.12), the control signal u is proposed:

$$u = K^{*T}X(t) + K_e e - \rho(\lambda B)^{-1}\frac{s}{\|s\|} \tag{10.13}$$

where $K^{*T} = \begin{bmatrix} k_{11}^* & k_{21}^* & k_{31}^* & k_{41}^* & k_{51}^* & k_{61}^* \\ k_{12}^* & k_{22}^* & k_{32}^* & k_{42}^* & k_{52}^* & k_{62}^* \\ k_{13}^* & k_{23}^* & k_{33}^* & k_{43}^* & k_{53}^* & k_{63}^* \end{bmatrix}$, $\|\cdot\|$ is the Euclidean norm, s is a column vector as $s = [s_1 \ \ s_2 \ \ s_3]^T$ and $\frac{s}{\|s\|}$ is the sliding mode unit control signal, ρ is constant.

The adaptive version of control input is:

$$u(t) = K^T(t)X(t) + K_e e(t) - \rho(\lambda B)^{-1}\frac{s}{\|s\|} \tag{10.14}$$

where $K(t)$ is an estimate of K^*.

Define the estimation error as:

$$\widetilde{K}(t) = K(t) - K^{*} . \tag{10.15}$$

Substituting (10.15) into (10.14) yields:

$$\dot{X}(t) = A_m X(t) + B\widetilde{K}^T(t)X(t) + BK_e e + Bf_m + f_u - B\rho(\lambda B)^{-1}\frac{s}{\|s\|} . \tag{10.16}$$

Then, we have the tracking error equation:

$$\dot{e}(t) = (A_m + BK_e)e + B\widetilde{K}^T(t)X(t) + Bf_m + f_u - B\rho(\lambda B)^{-1}\frac{s}{\|s\|} \tag{10.17}$$

and the derivative of $s(t)$ is:

$$\dot{s}(t) = \lambda B\widetilde{K}^T(t)X(t) + \lambda Bf_m + \lambda f_u - \rho\frac{s}{\|s\|} \cdot \tag{10.18}$$

Define a Lyapunov function:

$$V = \frac{1}{2}s^T s + \frac{1}{2}tr[\widetilde{K}M^{-1}\widetilde{K}^T] \tag{10.19}$$

where $M = M^T > 0$, M is positive definite matrix, $tr[M]$ denoting the trace of a square matrix M.

Differentiating V, with respect to time, yields:

$$\begin{aligned}\dot{V} &= s^T\dot{s} + tr[\widetilde{K}M^{-1}\dot{\widetilde{K}}^T] \\ &= s^T\left[\lambda B\widetilde{K}^T(t)X(t) + \lambda Bf_m + \lambda f_u - \rho\frac{s}{\|s\|}\right] + tr[\widetilde{K}M^{-1}\dot{\widetilde{K}}^T] \\ &= -s^T\rho\frac{s}{\|s\|} + s^T\lambda Bf_m + s^T\lambda f_u + s^T\lambda B\widetilde{K}^T(t)X(t) + tr[\widetilde{K}M^{-1}\dot{\widetilde{K}}^T].\end{aligned} \tag{10.20}$$

To make $\dot{V} \le 0$, we choose the adaptive law as:

$$\dot{\tilde{K}}^T(t) = \dot{K}^T(t) = -MB^T\lambda^T sX^T(t) \tag{10.21}$$

with $K(0)$ being arbitrary. This adaptive law yields:

$$\begin{aligned} \dot{V} &= -\rho\|s\| + s^T\lambda Bf_m + s^T\lambda f_u \\ &\le -\rho\|s\| + \|s\|\|\lambda B\|\|f_m\| + \|s\|\|\lambda\|\|f_u\| \\ &\le -\rho\|s\| + \|s\|\|\lambda B\|(\alpha_{m1}\|X\| + \alpha_{m2}) + \|s\|\|\lambda\|(\alpha_{u1}\|X\| + \alpha_{u2}) \\ &= -\|s\|[\rho - \|\lambda B\|(\alpha_{m1}\|X\| + \alpha_{m2}) - \|\lambda\|(\alpha_{u1}\|X\| + \alpha_{u2})] \le 0 \end{aligned} \tag{10.22}$$

with $\rho \ge \|\lambda B\|(\alpha_{m1}\|X\| + \alpha_{m2}) + \|\lambda\|(\alpha_{u1}\|X\| + \alpha_{u2}) + \eta$, where η is a positive constant, $\dot{V}$ becomes negative semi-definite, i.e., $\dot{V} \le -\eta\|s\|$. This implies that the trajectory reaches the sliding surface in finite time and remains on the sliding surface. The fact that $\dot{V}$ is negative semi-definite ensures V, s and $\tilde{K}$ are all bounded. It can be concluded from (10.18) that $\dot{s}$ is also bounded. The inequality $\dot{V} \le -\eta\|s\|$ implies that s is integrable as $\int_0^t\|s\|dt \le \frac{1}{\eta}[V(0) - V(t)]$. Since $V(0)$ is bounded and $V(t)$ is nonincreasing and bounded, it can be concluded that $\lim_{t\to\infty}\int_0^t\|s\|dt$ is bounded. Since $\lim_{t\to\infty}\int_0^t\|s\|dt$ is bounded and $\dot{s}$ is also bounded, according to Barbalat lemma, $s(t)$ will asymptotically converge to zero, $\lim_{t\to\infty} s(t) = 0$. From the adaptive laws (10.21), according to the persistence excitation theory, if X is the persistent excitation signal, then $\dot{\tilde{K}}^T(t) = -MB^T\lambda^T sX^T$ guarantees that $\tilde{K} \to 0$, K will converge to its true value. It can be shown that if $w_1 \ne w_2 \ne w_3$, there will always exist some positive scalar constants α and T, such that for all

$t > 0$, $\int_t^{t+T} XX^T d\tau \geq \alpha I$, where $XX^T = \begin{bmatrix} x_1^2 & x_1\dot{x}_1 & x_1x_2 & x_1\dot{x}_2 & x_1x_3 & x_1\dot{x}_3 \\ \dot{x}_1x_1 & \dot{x}_1^2 & \dot{x}_1x_2 & \dot{x}_1\dot{x}_2 & \dot{x}_1x_3 & \dot{x}_1\dot{x}_3 \\ x_2x_1 & x_2\dot{x}_1 & x_2^2 & x_2\dot{x}_2 & x_2x_3 & x_2\dot{x}_3 \\ \dot{x}_2x_1 & \dot{x}_2\dot{x}_1 & \dot{x}_2x_2 & \dot{x}_2^2 & \dot{x}_2x_3 & \dot{x}_2\dot{x}_3 \\ x_3x_1 & x_3\dot{x}_1 & x_3x_2 & x_3\dot{x}_2 & x_3^2 & x_3\dot{x}_3 \\ \dot{x}_3x_1 & \dot{x}_3\dot{x}_1 & \dot{x}_3x_2 & \dot{x}_3\dot{x}_2 & \dot{x}_3x_3 & \dot{x}_3^2 \end{bmatrix}$

.

It can be shown that XX^T has full rank if $w_1 \neq w_2 \neq w_3$, i.e. the excitation frequencies on x and y axes should be different. In other words, excitation of the proof mass should be persistently exciting. Since $\widetilde{K} \to 0$, then the unknown angular velocity, as well as all other unknown parameters, can be determined from $A + BK^T = A_m$.

In summary, if persistently exciting drive signals, $x_m = A_1 \sin(w_1 t)$, $y_m = A_2 \sin(w_2 t)$ and $z_m = A_3 \sin(w_3 t)$ are used, then $\widetilde{K}$ and $s(t)$ asymptotically converge to zero. Consequently, the unknown angular velocity can be determined as $\lim_{t\to\infty} \Omega_x(t) = \Omega_x$, $\lim_{t\to\infty} \Omega_y(t) = \Omega_y$ and $\lim_{t\to\infty} \Omega_z(t) = \Omega_z$. Nonetheless, it is difficult to establish the convergence rate.

Remark 2. To enable all unknown parameter estimates to converge to their true value, the necessary model trajectory is shown to be a 3D Lissajous pattern. The proposed device includes a single suspended mass that can move in three axes, having actuation and sensing elements in three orthogonal axes.

Remark 3. If (A_m, B) is a controllable pair, the closed-loop matrix $A_m + BK_e$ can have an arbitrary set of eigenvalues through the pole assignment method, obtaining the desired dynamics. . Through this process, tracking error settling rates can be controlled.

Remark 4. In order to eliminate the chattering, the discontinuous control component in (10.14) can be replaced by a smooth sliding mode component to yield:

$$u(t) = K^T(t)X(t) + K_e e - (\lambda B)^{-1} \rho \frac{s}{\|s\| + \varepsilon} \tag{10.23}$$

where $\varepsilon > 0$ is a small constant. This creates a small boundary layer about the switching surfaces in which the system trajectory will remain. Therefore, the chattering problem can be reduced significantly.

10.2. Simulation Example

We will evaluate the proposed adaptive sliding mode control on the lumped MEMS gyroscope model. For the convenience of simulation, only the matched uncertainty and distance are considered in the simulation section, allowing ± 5% parameter variations for the spring and damping coefficients and further assuming ± 5% magnitude changes in the coupling terms. These parameter variations constitute ΔA, assumed to have only match uncertainty. The external disturbance is a random variable with zero mean and unit variance, assumed to have only matched disturbance in the simulation.

Parameters of the MEMS triaxial gyroscope are as follows:

$m = 0.57e-8$ kg, $w_0 = 3kHz$, $q_0 = 10^{-6}m$. $d_{xx} = 0.429e-6$ N s/m, $d_{yy} = 0.687e-36$ N s/m, $d_{zz} = 0.895e-6$ N s/m, $d_{xy} = 0.0429e-6$ N s/m, $d_{xz} = 0.0687e-6$ N s/m, $d_{yz} = 0.0895e-6$ N s/m, $k_{xx} = 80.98$ N/m, $k_{xy} = 5$ N/m, $k_{yy} = 71.62$ N/m, $k_{zz} = 60.97$ N/m, $k_{xz} = 6$ N/m, $k_{yz} = 7$ N/m.

The unknown angular velocity is assumed as $\Omega_z = 5.0$ rad/s, $\Omega_x = 3.0$ rad/s, $\Omega_y = 2.0$ rad/s, and the initial condition of K is $K(0) = 0.9K^*$. The desired motion trajectories are $x_m = \sin(w_1 t)$, $y_m = 1.2\sin(w_2 t)$ and $z_m = 1.5\sin(w_3 t)$ where $w_1 = 6.71kHz$, $w_2 = 5.11kHz$ and $w_3 = 4.17kHz$. Figure 10.2 depicts the Lssajous trajectory of the reference model when $t = 10s$. The sliding mode parameter λ is chosen as: $\lambda = \begin{bmatrix} 0 & 10 & 0 & 0 & 0 & 0 \\ 0 & 0 & 0 & 10 & 0 & 0 \\ 0 & 0 & 0 & 0 & 0 & 10 \end{bmatrix}$,

and K_e is designed as:

$$K_e = \begin{bmatrix} -10000 & -10000 & -10000 & -10000 & 0 & 10000 \\ -10000 & -10000 & -10000000 & -10000 & 10000 & 10000 \\ -10000 & -10000 & -100000000 & -10000 & 10000 & -10000 \end{bmatrix}.$$

The adaptive gain is $M = diag\{20 \quad 20 \quad 20\}$, and the sliding mode gain is $\rho = diag\{10000 \quad 10000 \quad 10000\}$, $\varepsilon = 0.15$.

The convergence behavior of sliding surface and tracking errors are shown in Figures 10.3-4. It can be shown that $s(t)$ and $e(t)$ will asymptotically converge to zero. Figure 10.5 and Figure 10.6 compare the angular velocity estimation between the smooth sliding mode controller and non-smooth sliding mode controller. It can be observed that both Figures 5 and 8 illustrate that the estimates of angular velocity converge to their true values when with a persistent excitation signal, $\lim_{t\to\infty}\hat{\Omega}_x(t)=\Omega_x$, $\lim_{t\to\infty}\hat{\Omega}_y(t)=\Omega_y$ and $\lim_{t\to\infty}\hat{\Omega}_z(t)=\Omega_z$.

However, the estimated angular velocities with smooth sliding mode controller have better convergence performance. The true value of the controller is:

$$K^* = \begin{bmatrix} 1533.5 & 0.0251 & 97.4659 & 9.9975 & 116.9591 & 4.0025 \\ 97.4659 & 10.004 & 1370 & 0.0402 & 136.4522 & 5.9948 \\ 116.9591 & 3.996 & 136.4522 & 6.0052 & 1167.6 & 0.0523 \end{bmatrix};$$

therefore, it can be demonstrated from Figure 10.7 that the estimates of controller parameters converge to their true values with a persistent excitation signal. Figure 10.8 depicts the sliding mode control force with a smooth sliding mode controller. It is shown that the adaptive control system with the smooth sliding mode controller can significantly reduce chattering That the smooth sliding mode controller creates a small boundary layer about the switching surface is the reason why a smooth sliding mode component can reduce the chattering problem.

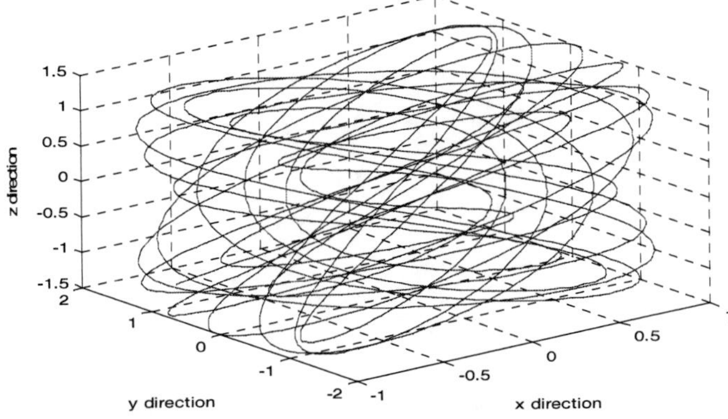

Figure 10.2. Reference model Lssajous trajectory.

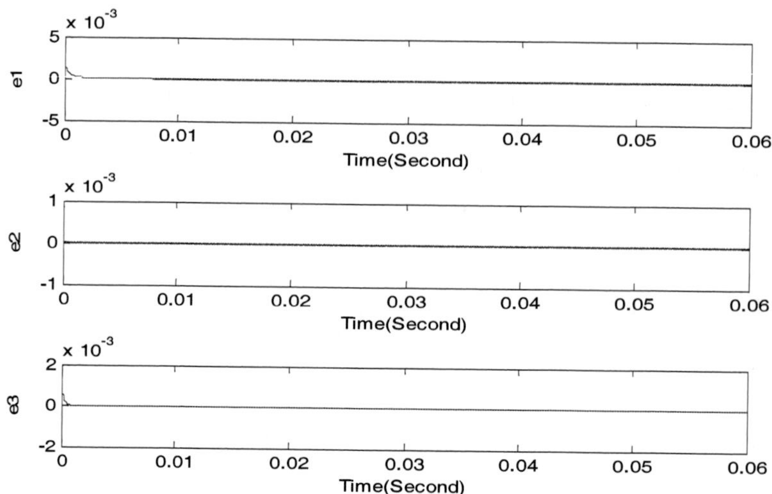

Figure 10.3. Convergence of the tracking error e(t).

The system response is as expected: with the control law (10.14) and the parameter adaptation laws (10.21), if the gyroscope is controlled to follow the mode-unmatched reference model with $w_1 \neq w_2 \neq w_3$, in other words, persistently exciting drive signals, $x_m = A_1 \sin(w_1 t)$, $y_m = A_2 \sin(w_2 t)$ and $z_m = A_3 \sin(w_3 t)$ are used, all unknown gyroscope parameters including the angular velocities converge to their true values and $\tilde{\theta}(t)$, $s(t)$ and $e(t)$ are going to zero asymptotically.

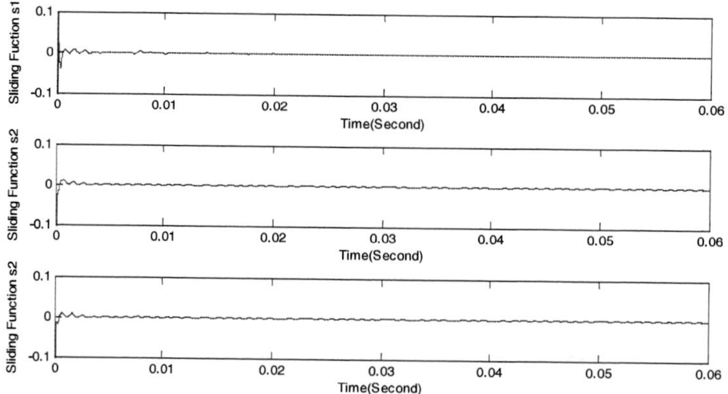

Figure 10.4. Convergence of the sliding surface s(t).

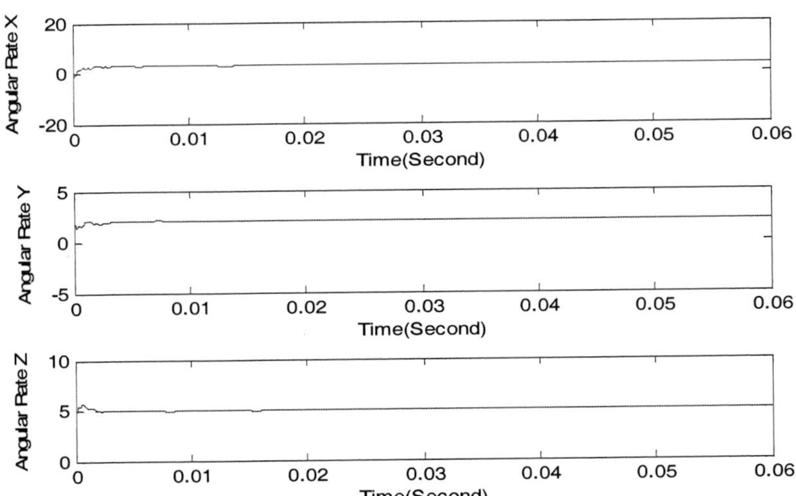

Figure 10.5 Convergence of angular velocity using smooth sliding mode controller (True values: $\Omega_x = 3.0$ rad/s, $\Omega_y = 2.0$ rad/s, $\Omega_z = 5.0$ rad/s).

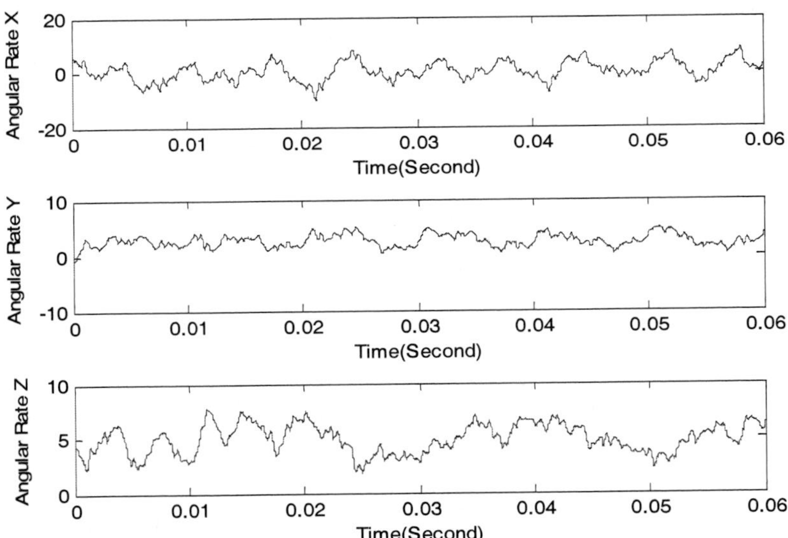

Figure 10.6. Convergence of angular velocity using non-smooth sliding mode controller (True values $\Omega_x = 3.0$ rad/s, $\Omega_y = 2.0$ rad/s, $\Omega_z = 5.0$ rad/s).

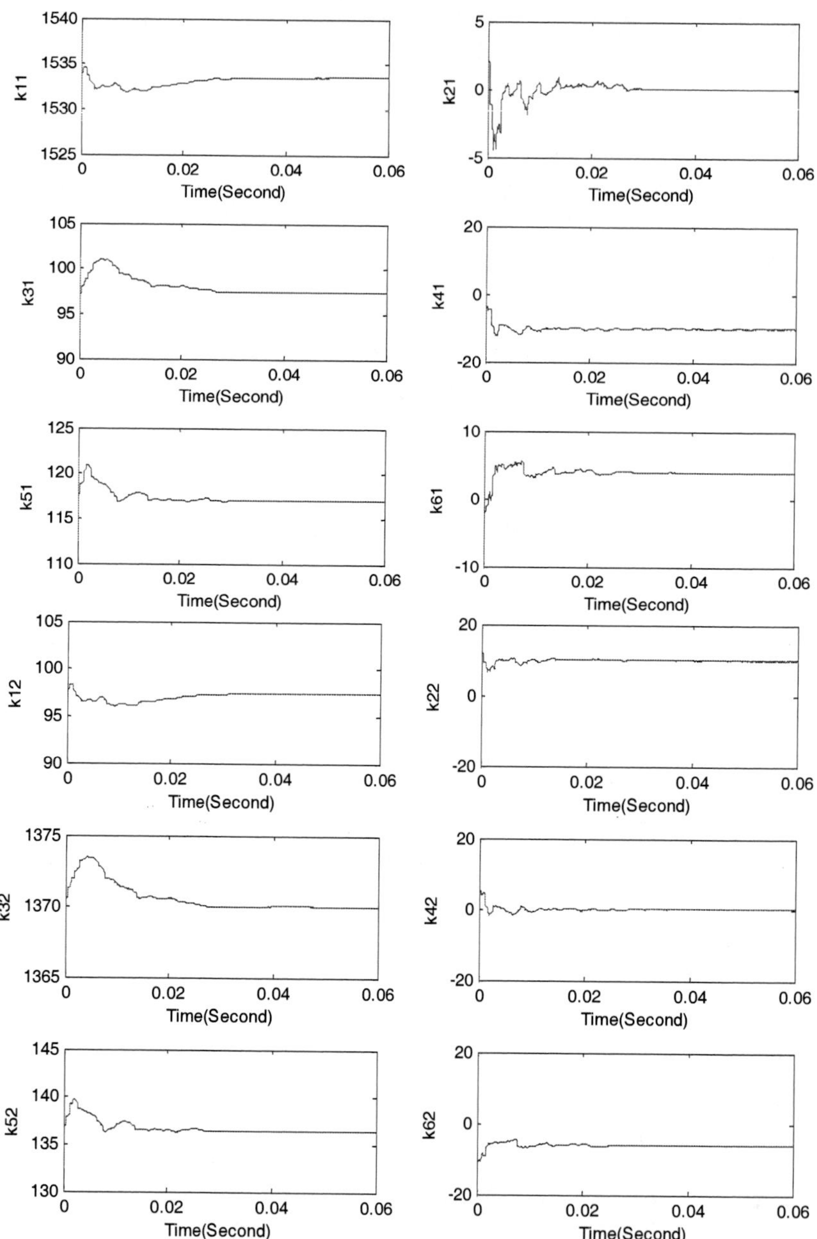

Figure 10.7. (Continued).

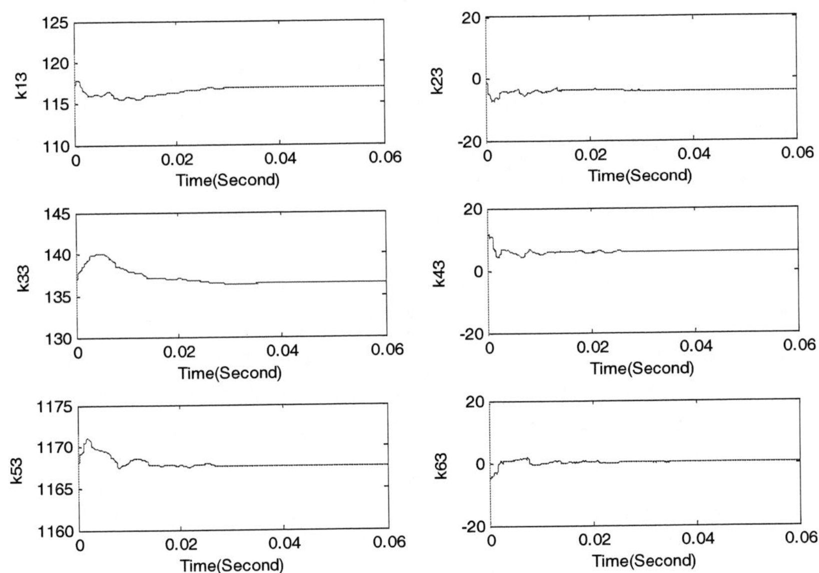

Figure 10.7. Adaptation of the controller parameters.

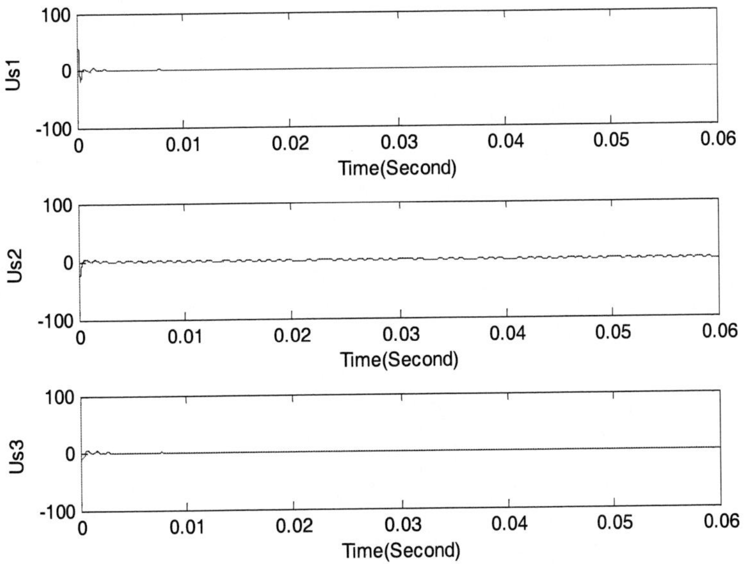

Figure 10.8. Smooth sliding mode control force of the adaptive sliding mode controller.

CONCLUDING REMARKS

This chapter presented an adaptive sliding mode control for the state tracking control of a triaxial angular velocity sensor. A triaxial sensor substrate was adapted for use in measuring the acceleration and angular velocity of a moving body along three orthogonal axes. The controller drives the single mass along a controlled oscillation trajectory, removing the need for additional drive control. To enable all unknown parameters estimated to converge to their true values, the necessary model trajectory was shown to be a 3D Lissajous pattern. The proposed device included a single suspended mass that can move in three axles, having actuation and sensing elements in three orthogonal axes. The simulation is implemented to verify the effectiveness of the proposed adaptive sliding mode control for this triaxial angular velocity sensor.

Chapter 11

Neural Network Based Adaptive Sliding Mode Control of MEMS Gyroscopes

Using a radial basis function (RBF) network, this chapter presents a robust adaptive sliding mode control strategy of MEMS triaixal gyroscope. A key property of this scheme is that a prior knowledge of the upper bound of the system uncertainties is not required. An adaptive RBF neural network is used to learn the unknown upper bound of model uncertainties and external disturbances. In addition, the proposed adaptive sliding mode controller updates estimates of all stiffness errors, damping terms and angular velocities in real time. The adaptive RBF neural network is incorporated into the adaptive sliding mode control scheme in the same Lyapunov framework and the stability of the proposed adaptive neural sliding mode control can be established. The numerical simulation for a MEMS triaxial angular velocity sensor is investigated to verify the effectiveness of the proposed adaptive sliding mode control scheme, as based on a RBF network.

In the last few years, some control algorithms have been proposed to control the MEMS gyroscope. This chapter develops adaptive neural network control to estimate the upper bound of system uncertainties. Intelligent control approaches, such as neural network and fuzzy control, do not require mathematical models and have ability to approximate nonlinear systems. Guo and Woo [90] and Wai [91] proposed adaptive fuzzy sliding mode controllers for robot manipulators. Recently, neural network technology has been applied into the control system [92-98] and the neural network's learning ability to approximate arbitrary nonlinear functions, making it a useful tool for adaptive application.

The control scheme presented in this chapter integrates the theory of adaptive sliding mode control and the nonlinear mapping of neural network. A RBF neural network is used to adaptively learn the unknown bounds of system uncertainties. The chattering of a sliding mode can effectively be eliminated. The control laws could guarantee a convergence of the trajectory tracking error and unknown system parameters, as well as robustness for external disturbances and model uncertainties. Meanwhile, the proposed adaptive sliding mode controller updates estimates of all stiffness errors, damping and angular velocities in real time.

11.1. Adaptive Sliding Mode Control Using a RBF Network

In this section, because of the great advantages of neural networks in dealing with the nonlinear system, an adaptive neural sliding mode controller is designed and the stability of the proposed adaptive control with adaptive neural sliding mode control is analyzed. The block diagram of an adaptive sliding mode control of triaxial gyroscopes using a RBF network is designed, as shown in Figure 11.1.

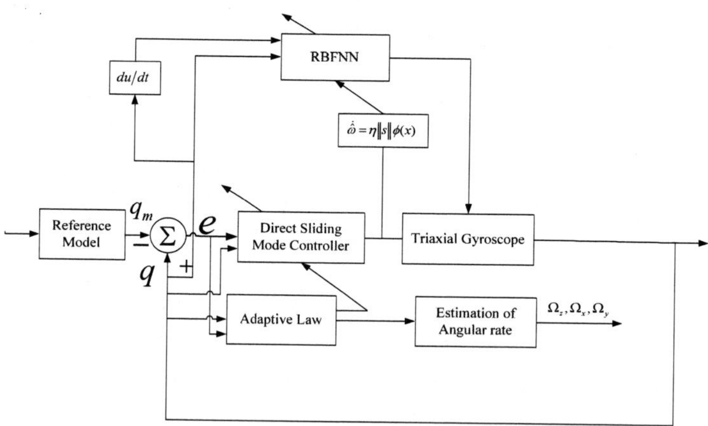

Figure 11.1. Block diagram of adaptive sliding mode control using RBF network.

The control target for a MEMS gyroscope is to maintain the proof mass to oscillate in the x , y and z directions at a given frequency and amplitude:

$$x_m = A_1 \sin(\omega_1 t), y_m = A_2 \sin(\omega_2 t), z_m = A_3 \sin(\omega_3 t) .$$

The reference model is defined as:

$$\ddot{q}_m + K_m q_m = 0 \tag{11.1}$$

where $K_m = diag\{\omega_1^{\ 2}, \omega_2^{\ 2}, \omega_3^{\ 2}\}$.

Consider the dynamics with parametric uncertainties and external disturbance as:

$$\ddot{q} + (D + 2\Omega + \Delta D)\dot{q} + (K_b + \Delta K_b)q = u + d \tag{11.2}$$

where ΔD is unknown uncertainties of the matrix $D + 2\Omega$, ΔK_b is the unknown parameter uncertainties of the matrix K_b , and d is an uncertain extraneous disturbance of the system .

Rewriting (11.2) as:

$$\ddot{q} + (D + 2\Omega)\dot{q} + K_b q = u + \rho \tag{11.3}$$

where ρ represents the lumped model uncertainties and external disturbances, given by:

$$\rho = d - \Delta D\dot{q} - \Delta K_b q . \tag{11.4}$$

Define:

$$K_3^{*T} = K_m - K_b, K_4^{*T} = -\left(D + 2\Omega\right). \tag{11.5}$$

where

$$K_3^* = \begin{bmatrix} k_{11}^* & k_{12}^* & k_{13}^* \\ k_{21}^* & k_{22}^* & k_{23}^* \\ k_{31}^* & k_{32}^* & k_{33}^* \end{bmatrix} = \begin{bmatrix} w_1^2 - w_x^2 & -w_{xy} & -w_{xz} \\ -w_{xy} & w_2^2 - w_y^2 & -w_{yz} \\ -w_{xz} & -w_{yz} & w_3^2 - w_z^2 \end{bmatrix},$$

$$K_4^* = \begin{bmatrix} k_{41}^* & k_{42}^* & k_{43}^* \\ k_{51}^* & k_{52}^* & k_{53}^* \\ k_{61}^* & k_{62}^* & k_{63}^* \end{bmatrix} = -\begin{bmatrix} d_{xx} & d_{xy} - 2\Omega_z & d_{xz} + 2\Omega_y \\ d_{xy} + 2\Omega_z & d_{yy} & d_{yz} - 2\Omega_x \\ d_{xz} - 2\Omega_y & d_{yz} + 2\Omega_x & d_{zz} \end{bmatrix}$$

Substitute (11.5) into (11.3), then the system dynamics becomes:

$$\ddot{q} + K_m q = u + K_3^{*T} q + K_4^{*T} \dot{q} + \rho \quad . \tag{11.6}$$

The tracking error and its derivative are:

$$\begin{cases} e_1 = q - q_m \\ e_2 = \dot{q} - \dot{q}_m \end{cases} \tag{11.7}$$

where q_m and $\dot{q}_m$ are states of the reference model: (11.1), $e_1 = [e_{11} \quad e_{12} \quad e_{13}]^T, e_2 = [\dot{e}_{11} \quad \dot{e}_{12} \quad \dot{e}_{13}]^T$, $e_{11} = x(t) - x_m(t), e_{12} = y(t) - y_m(t), e_{13} = z(t) - z_m(t)$.

The derivative of tracking errors can be derived as:

$$\begin{cases} \dot{e}_1 = \dot{q} - \dot{q}_m = e_2 \\ \dot{e}_2 = u + K_3^{*T} q + K_4^{*T} \dot{q} + \rho - K_m q + K_m q_m \end{cases} \quad . \tag{11.8}$$

The sliding surface is defined as:

$$s = ce_1 + e_2 \tag{11.9}$$

where c is constant matrix.

The derivative of the sliding surface becomes:

$$\dot{s} = c\dot{e}_1 + \dot{e}_2 = ce_2 + \dot{e}_2 \quad . \tag{11.10}$$

Substituting (11.8) into (11.10) yields:

$$\dot{s} = ce_2 + u + K_3^{*T} q + K_4^{*T} \dot{q} + \rho - K_m q + K_m q_m \ . \qquad (11.11a)$$

Setting $\dot{s} = 0$ to solve equivalent control u_{eq} gives:

$$u_{eq} = -ce_2 - K_3^{*T} q - K_4^{*T}(t)\dot{q} + \rho - K_m q + K_m q_m \ . \qquad (11.11b)$$

An adaptive version of control algorithm is proposed as:

$$u(t) = -ce_2 - K_3^T(t)q - K_4^T(t)\dot{q} - K_m q + K_m q_m - \hat{\bar{\rho}}(t)\frac{s}{\|s\|} \qquad (11.12)$$

where $s = [s_1 \quad s_2 \quad s_3]^T$, $\hat{\bar{\rho}}(t)$ is the variable matrix, and $\frac{s}{\|s\|}$ is the sliding mode unit control signal, $K_3\ (t)$ and $K_4(t)$ is the estimate of K_3^* and K_4^*.

Define the estimation errors as:

$$\tilde{K}_3(t) = K_3^* - K_3(t),\ \tilde{K}_4(t) = K_4^* - K_4(t) \ . \qquad (11.13)$$

The dynamics of sliding surface becomes:

$$\dot{s}(t) = \tilde{K}_3^T(t)q + \tilde{K}_4^T(t)\dot{q} - \hat{\bar{\rho}}(t)\frac{s}{\|s\|} + \rho(t) \ . \qquad (11.14)$$

Suppose $\bar{\rho}(t)$ is the upper bound of the end $|\rho(t)| < \bar{\rho}(t)$. If the upper bound value $\bar{\rho}(t)$ can not be measured properly and is unknown, the RBF neural network can be used to adaptively learn the upper bound $\bar{\rho}(t)$. The structure of RBF neural networks is a three-layer feedforward, as shown in Figure 11.2. The input layer is the set of source modes. The second layer is a hidden layer of high dimension. The output layer gives the response of the network to the activation patterns, as applied to the input layer. The transformation from the input space to the hidden-unit space is nonlinear, while the transformation from the hidden space to the output space is linear. Furthermore, from input to output the mapping is nonlinear, and thus greatly accelerates the learning speed and avoids a local minimum problem. In this chapter, the advantage of the RBF neural network is to adjust the value of the upper bound of uncertain systems.

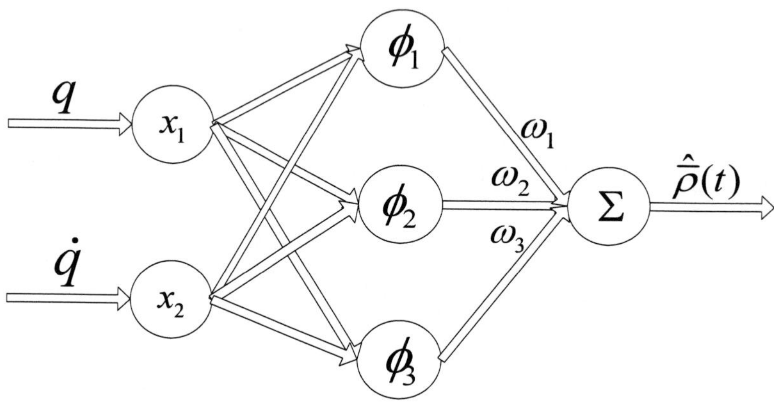

Figure 11.2. The structure of a RBF network.

The estimate of the upper bound $\bar{\rho}(t)$ is $\hat{\bar{\rho}}(x,\omega) = \hat{\omega}^T \phi(x)$ (11.15)

where $x = [q \quad \dot{q}]$ is the input of a RBF neural network, $\hat{\omega}^T$ are weights of a RBF neural network and $\phi(x)$ is Gaussion function:

$$\phi_i(x) = \exp(-\frac{\|x - m_i\|^2}{\sigma_i^2}), i = 1, \quad 2, \quad 3 \tag{11.16}$$

where m_i is the center of number i neurons, σ_i is width of number i neurons.

We make the following assumptions:

A1: The optimal weights of the RBF satisfy:

$$\omega^{*T} \phi(x) - \bar{\rho}(t) = \varepsilon(x) \text{ and } |\varepsilon(x)| < \varepsilon_1 \tag{11.17}$$

A2: The upper bound $\bar{\rho}(t)$ satisfies:

$$\bar{\rho}(t) - |\rho(t)| > \varepsilon_0 > \varepsilon_1 \tag{11.18}$$

Define a Lyapunov function:

$$V = \frac{1}{2}s^T s + \frac{1}{2}tr[\tilde{K}_3 M^{-1}\tilde{K}_3{}^T + \tilde{K}_4 M^{-1}\tilde{K}_4{}^T] + \frac{1}{2}\eta^{-1}\tilde{\omega}^T\tilde{\omega} \tag{11.19}$$

where $\tilde{\omega} = \omega^* - \hat{\omega}$, $\eta = \varepsilon_0 - \varepsilon_1 > 0$, $M = M^T > 0$, M is positive definite matrix, tr[M] denoting the trace of M .

The derivative of the Lyapunov function is:

$$\begin{aligned}
\dot{V} &= s^T\dot{s} + tr[\tilde{K}_3 M^{-1}\dot{\tilde{K}}_3{}^T + \tilde{K}_4 M^{-1}\dot{\tilde{K}}_4{}^T] - \eta^{-1}\tilde{\omega}^T\dot{\hat{\omega}} \\
&= s^T s + [\tilde{k}_3{}^T(t)q + \tilde{k}_4{}^T(t)\dot{q} - \hat{\bar{\rho}}(t)\frac{s}{\|s\|} + \rho(t)] \\
&\quad + tr[\tilde{k}_3 M^{-1}\dot{\tilde{k}}_3{}^T + \tilde{k}_4 M^{-1}\dot{\tilde{k}}_4{}^T] - \eta^{-1}\tilde{\omega}^T\dot{\hat{\omega}} \\
&= s^T\tilde{K}_3{}^T(t)q + s^T\tilde{K}_4{}^T(t)\dot{q} + tr[\tilde{K}_3 M^{-1}\dot{\tilde{K}}_3{}^T + \tilde{K}_4 M^{-1}\dot{\tilde{K}}_4{}^T] \\
&\quad + s^T[\rho(t) - \hat{\bar{\rho}}(t)\frac{s}{\|s\|}] - \eta^{-1}\tilde{\omega}^T\dot{\hat{\omega}}
\end{aligned} \tag{11.20}$$

Making using of properties of trace yields:

$$s^T\tilde{K}_3{}^T q = tr[qs^T\tilde{K}_3{}^T(t)] = tr[\tilde{K}_3(t)sq^T],$$

$$s^T\tilde{K}_4{}^T\dot{q} = tr[\dot{q}s^T\tilde{K}_4{}^T(t)] = tr[\tilde{K}_4(t)s\dot{q}^T] \tag{11.21}$$

Then, (11.20) becomes:

$$\begin{aligned}
\dot{V} &= tr[\tilde{K}_3(t)sq^T] + tr[\tilde{K}_4(t)s\dot{q}^T] + tr[\tilde{K}_3 M^{-1}\dot{\tilde{K}}_3{}^T + \tilde{K}_4 M^{-1}\dot{\tilde{K}}_4] \\
&\quad + s^T[\rho(t) - \hat{\bar{\rho}}(t)\frac{s}{\|s\|}] - \eta^{-1}\tilde{\omega}^T\dot{\hat{\omega}}
\end{aligned} \tag{11.22}$$

To make $\dot{V} \le 0$,we choose the adaptive laws as:

$$\begin{aligned} \dot{\tilde{K}}_3^T(t) &= -Msq^T \\ \dot{\tilde{K}}_4^T(t) &= -Ms\dot{q}^T \end{aligned} \tag{11.23}$$

Substituting $\dot{\tilde{K}}_3^T(t), \dot{\tilde{K}}_4^T(t)$ into $\dot{V}$ yields:

$$\begin{aligned} &\dot{V} = s^T[\rho(t) + \bar{\rho}(t) - \bar{\rho}(t) - \hat{\rho}(t)\frac{s}{\|s\|}] - \eta^{-1}\tilde{\omega}^T\dot{\hat{\omega}} \\ &\le \|s\|\left[|\rho(t)| + \bar{\rho}(t) - \bar{\rho}(t)\right] - \hat{\rho}(t)\|s\| - \eta^{-1}\tilde{\omega}^T\dot{\hat{\omega}} \\ &= -\|s\|\left[\bar{\rho}(t) - |\rho(t)|\right] + \left[\bar{\rho}(t)\|s\| - \hat{\rho}(t)\|s\|\right] - \eta^{-1}\tilde{\omega}^T\dot{\hat{\omega}} \\ &= -\|s\|\left[\bar{\rho}(t) - |\rho(t)|\right] + \|s\|\left[\omega^{*T}\phi(x) + \varepsilon(x) - \hat{\omega}^T\phi(x)\right] - \eta^{-1}\tilde{\omega}^T\dot{\hat{\omega}} \\ &= -\|s\|\left[\bar{\rho}(t) - |\rho(t)|\right] + \|s\|\varepsilon(x) + \|s\|\tilde{\omega}^T\phi(x) - \eta^{-1}\tilde{\omega}^T\dot{\hat{\omega}} \end{aligned} \tag{11.24}$$

Using an adaptive algorithm to adjust the weights online:

$$\dot{\hat{\omega}} = \eta\|s\|\phi(x) \tag{11.25}$$

Substituting $\dot{\hat{\omega}}$ into $\dot{V}$ yields:

$$\begin{aligned} &\dot{V} == -\|s\|\left[\bar{\rho}(t) - |\rho(t)|\right] + \|s\|\varepsilon(x) \le -\|s\|\varepsilon_0 + \|s\|\varepsilon \\ &\le -\|s\|\varepsilon_0 + \|s\|\varepsilon_1 = -\|s\|(\varepsilon_0 - \varepsilon_1) = -\eta\|s\| \le 0 \end{aligned} \tag{11.26}$$

$\dot{V}$ becomes negative semi-definite, implying that the trajectory reaches the sliding surface in finite time and remains on the sliding surface. $\dot{V}$ is negative definite implies that s , $\tilde{K}_3$ and $\tilde{K}_4$ converge to zero. That $\dot{V}$ is negative semi-definite ensures that V , s , $\tilde{K}_3$ and $\tilde{K}_4$ are all bounded. It can be concluded, from (11.14) that $\dot{s}$ is also bounded. The inequality of (11.28) implies that s is integrable as $\int_0^t \|s\| dt \le \frac{1}{\eta}[V(0) - V(t)]$. Since $V(0)$ is bounded and $V(t)$ is nonincreasing and bounded, it can be concluded that $\lim_{t\to\infty}\int_0^t \|s\| dt$ is bounded. Since $\lim_{t\to\infty}\int_0^t \|s\| dt$ is bounded and $\dot{s}$ is also bounded, according to Barbalat lemma, $s(t)$ will asymptotically converge to zero, $\lim_{t\to\infty} s(t) = 0$. As seen from (11.11), $e(t)$ also converges to zero asymptotically. From the adaptive laws (11.23), according to the persistence excitation theory, if q and $\dot{q}$ are persistent excitation signals, i.e. $\omega_1 \neq \omega_2 \neq \omega_3$, then it can be guaranteed that $\widetilde{K}_3 \to 0$, and $\widetilde{K}_4 \to 0$, K_3 and K_4 will converge to their true values.

11.2. Simulation Study

In the simulation, the parameters of the MEMS triaxial gyroscope are:

$m = 0.57e - 8kg, \omega_0 = 3kHz, q_0 = 10^{-6} m, d_{xx} = 0.429e - 6Ns/m, d_{yy} = 0.0429e - 6Ns/m,$

$d_{zz} = 0.895e - 6Ns/m, d_{xy} = 0.0429e - 6Ns/m, d_{xz} = 0.0687e - 6Ns/m, d_{yz} = 0.0895e - 6Ns/m,$

$k_{xx} = 80.98N/m, k_{xy} = 5N/m, k_{yy} = 71.62N/m, k_{zz} = 60.97N/m, k_{xz} = 6N/m, k_{yz} = 7N/m.$

The unknown angular velocity is assumed as:

$\Omega_z = 5.0rad/s, \Omega_x = 3.0rad/s, \Omega_y = 2.0rad/s.$

The desired motion trajectories are:

$x_m = \sin(\omega_1 t), y_m = 1.2\sin(\omega_2 t), z_m = 1.5\sin(\omega_3 t)$

where $\omega_1 = 6.71kHz, \omega_2 = 5.11kHz, \omega_3 = 4.17kHz$.

In (11.12) and (11.23), the parameters are chosen as:

$$c = \begin{bmatrix} 40 & 0 & 0 \\ 0 & 40 & 0 \\ 0 & 0 & 40 \end{bmatrix}, M = \begin{bmatrix} 20 & 0 & 0 \\ 0 & 20 & 0 \\ 0 & 0 & 20 \end{bmatrix}.$$

The initial value of ω is $\begin{bmatrix}0.1 & 0.1 & 0.1\end{bmatrix}^T$,the initial value of m is always chosen between -1 and +1, we choose $m = \begin{bmatrix} -0.1639 & 0.7487 & 0.5359 \\ -0.3900 & -0.9700 & 0.9417 \end{bmatrix}$ and $\sigma = \begin{bmatrix}0.2 & 0.2 & 0.2\end{bmatrix}^T$.

Initial value of system's states is $\begin{bmatrix}0.5 & 0\end{bmatrix}$, external disturbance $d(t) = 0.5\sin(2\pi t)$. The initial values of $K_3(t)$ and $K_4(t)$ are

$$K_3 = \begin{bmatrix} 1380.15 & 87.7193 & 105.2632 \\ 87.7193 & 1233 & 122.8070 \\ 105.2632 & 122.8070 & 1050.84 \end{bmatrix}, K_4 = \begin{bmatrix} 0.0226 & 8.9978 & 3.6023 \\ 9.0036 & 0.0362 & 5.3953 \\ 3.5964 & 5.4047 & 0.0471 \end{bmatrix}$$

Figure 11.3 shows the position tracking of X, Y and Z. It can be observed that the positions of X, Y and Z can track the position of the reference model in very short time, demonstrating that the adaptive tracking performance is satisfactory. Figure 11.4 depicts the convergence of the tracking errors. It can be seen that tracking errors converge to zero asymptotically. It can be concluded from Figures 11.3 and 11.4 that the MEMS gyroscope can maintain the proof mass to oscillate in the x, y and z directions at a given frequency and amplitude. The convergence of sliding surfaces is shown in Figure 11.5. It can be found that s(t) asymptotically converges to zero. The sliding system arrives at the sliding surface in a short time. After that, the control system will get into sliding mode trajectory and remain with it. Adaptation of the controller parameters is described in Figure 11.6, showing that $K_3(t), K_4(t)$ can converge to stable values in a short time. The true values of the controller $K_3(t), K_4(t)$ are:

$$K_3^* = \begin{bmatrix} 1533.5 & 97.4659 & 116.9591 \\ 97.4659 & 1370 & 136.4522 \\ 116.9591 & 136.4522 & 1167.6 \end{bmatrix},$$

$$K_4^* = \begin{bmatrix} 0.0251 & 9.9975 & 4.0025 \\ 10.004 & 0.0402 & 5.9948 \\ 3.996 & 6.0052 & 0.0523 \end{bmatrix}.$$

Therefore, it can be shown from Figure 11.6 that the estimates of controller parameters converge to their true values with a persistent excitation signal. The upper bound of system disturbance, estimated by the RBF neural network, are drawn in Figure 11.7: the RBF neural network is used to adjust the gain of the switch part of adaptive sliding mode control input.

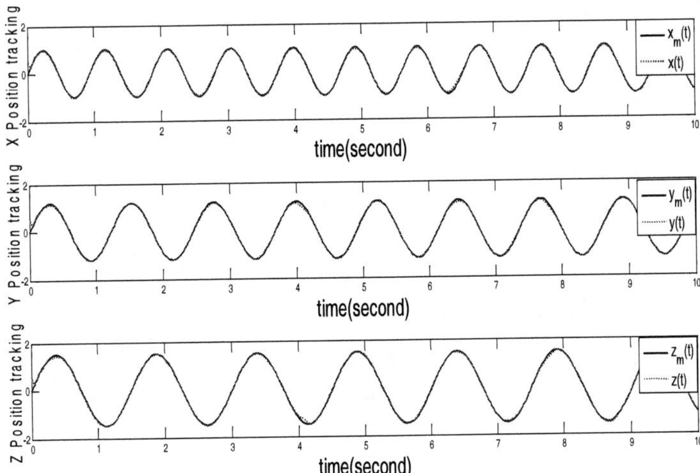

Figure 11.3. Position tracking of X ,Y and Z.

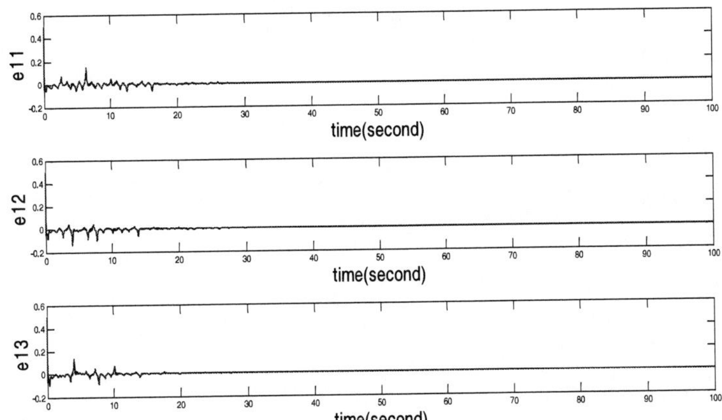

Figure 11.4. Convergence of the tracking error e(t).

Figure 11.8 is the adaptive sliding control input with adaptive upper bound of system disturbances, using the RBF neural network. Figure 11.9 is the adaptive sliding control input with fixed value upper bound of system disturbances.

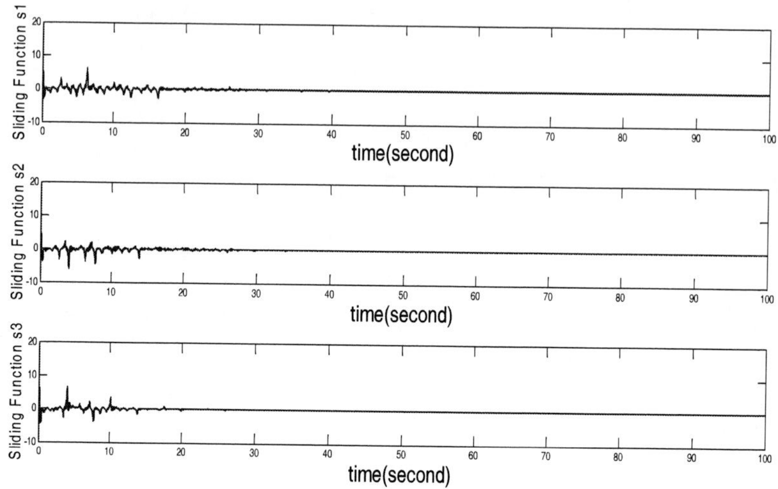

Figure 11.5. Convergence of the sliding surface s(t).

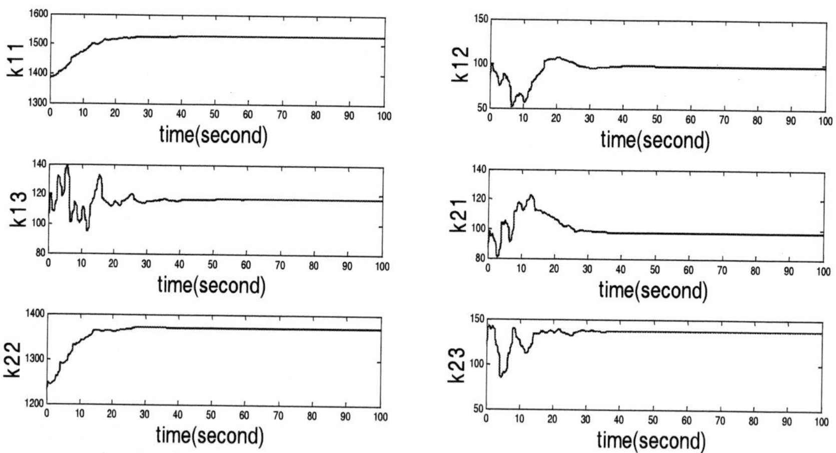

Figure 11.6. (Continued).

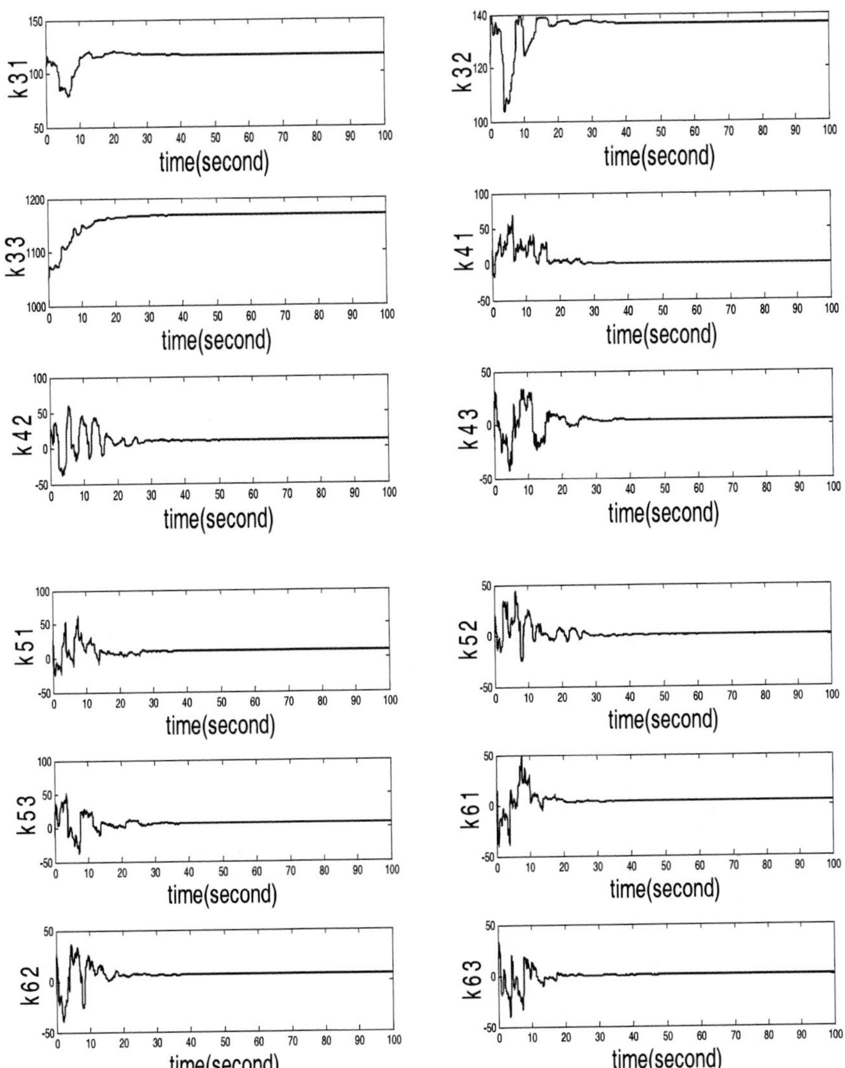

Figure 11.6. Adaptation of the controller parameters.

Comparing these two figures, it can be found that the control input in Figure 11.8 is better than that of Figure 11.9, and that the chattering problem is obvious when using a fixed value of upper bound of system disturbances. It is shown that the upper bound of adaptive learning by the RBF neural network can reduce chattering significantly.

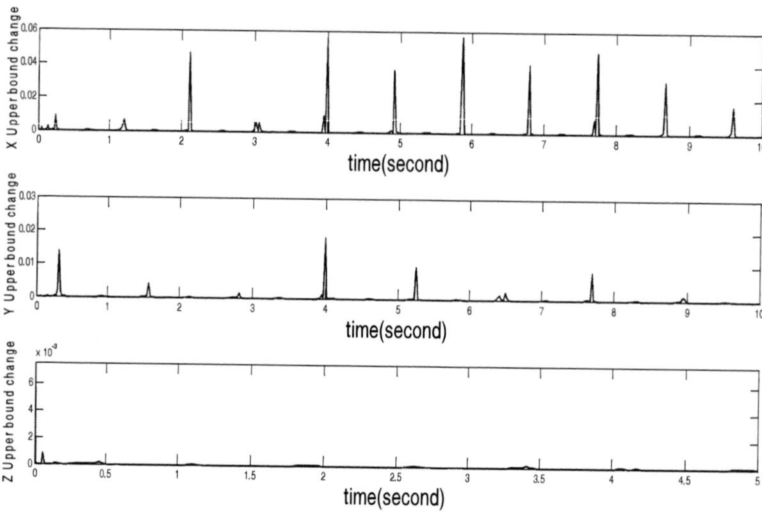

Figure 11.7. Adaptation of upper bound of system uncertainties.

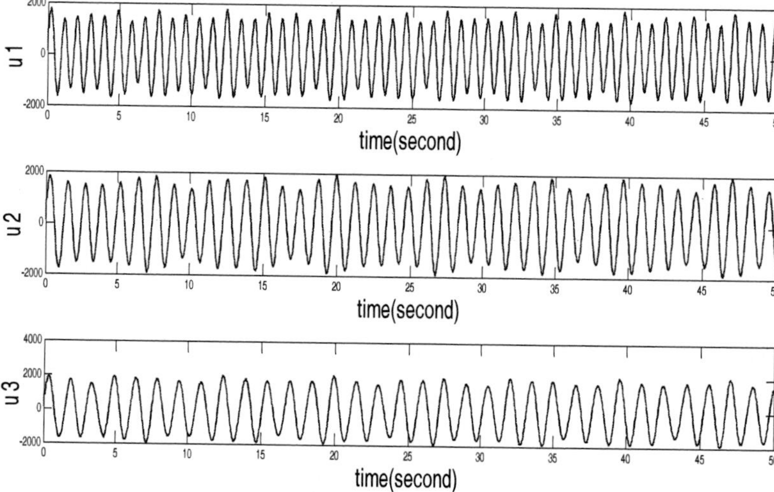

Figure 11.8. Control input with adaptive upper bound of system uncertainties using RBF network.

The RBF neural network could generate the smooth sliding mode control force, creating a small boundary layer about the switching surface to eliminate the chattering. Figure 11.10 depicts the angular velocity using a smooth sliding mode controller, showing that the estimates of angular velocity converge to their true values with persistent excitation signals. It can be concluded that the system response is as expected, and the simulation results show that the designed control system has a good tracking performance and a high control precision. Furthermore, system parameters, tracking errors and sliding surface converge to zero asymptotically.

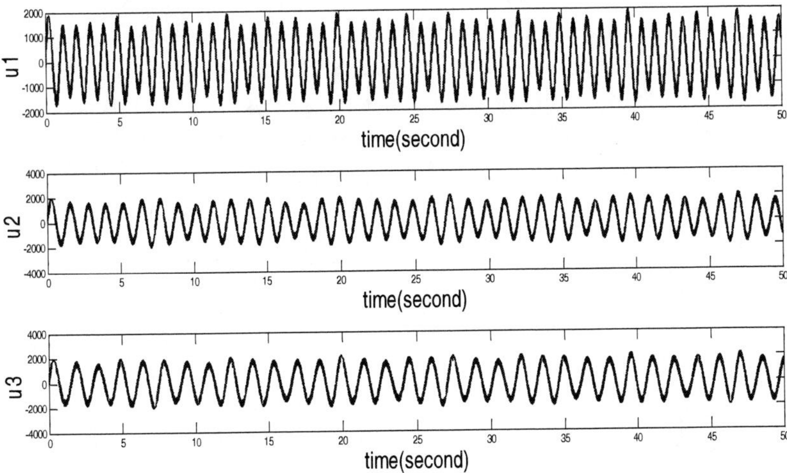

Figure 11.9. Control input using fixed value of upper bound of system uncertainties.

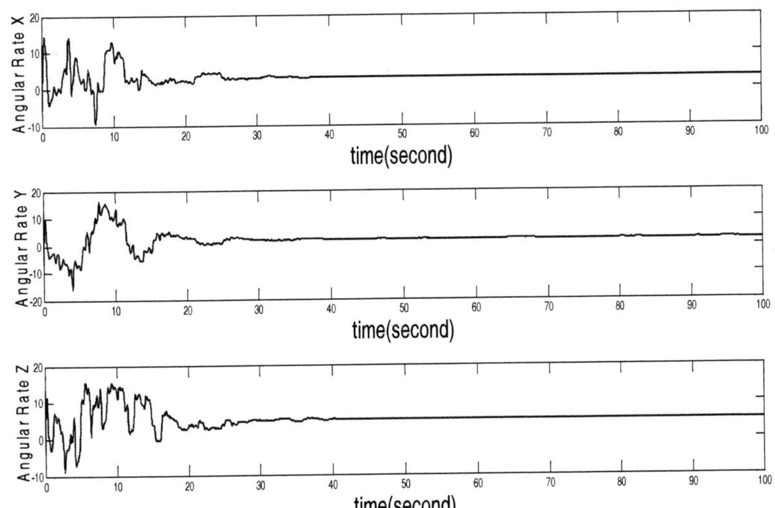

Figure 11.10. Convergence of angular velocity using smooth sliding mode controller.

CONCLUDING REMARKS

This chapter presented the RBF network based adaptive sliding mode control for the state tracking control of the triaxial angular velocity sensor. The controller drove the single mass along a controlled oscillation trajectory, removing the need for additional drive control. An adaptive RBF neural network was used to learn the upper bound of model uncertainties and external disturbances. The output of the neural network was used as a compensatory parameter and the effects of the system uncertainties could be eliminated. With the proposed adaptive RBF sliding mode control strategy, the stability of the closed-loop system could be guaranteed. The simulation was implemented to verify the effectiveness of the proposed adaptive RBF sliding mode control for this particular triaxial angular velocity sensor.

Chapter 12

CONCLUSIONS

This book developed advanced control strategies for a MEMS z-axis gyroscope. The book describes how control design moves from an adaptive control of the MEMS gyroscope to a robust adaptive control of the MEMS gyroscope. A direct adaptive sliding mode controller, with proportional and integral sliding surface for a time-varying linear system with single and multiple inputs, were also developed. The proposed adaptive sliding mode controllers for a MEMS z-axis gyroscope make real-time estimates of the angular velocity, as well as all other gyroscope parameters including coupling stiffness and damping parameters. Therefore, fabrication imperfection and time varying noise and disturbance can be compensated for. The reference model trajectory was designed to satisfy the persistent excitation and to enable the estimations of the parameters to converge to their true values.

A sliding mode control algorithm was incorporated into an adaptive control system and the feasibility of adaptive sliding mode control in the presence of the model uncertainty. External disturbance was also investigated in this book. The technique had been applied to a MEMS gyroscope and the angular velocity and all gyroscope parameters are consistently determined. A robust adaptive control scheme, without an integral sliding surface; and an adaptive control scheme, with proportional-integral sliding surface, were investigated. The proposed control scheme differs from the previous sliding mode techniques in the sense that the sliding surface is based on the proportional integral sliding mode control. The additional integral term provides one more degree of freedom than the conventional sliding surface. The advantage of the adaptive sliding mode controller with integral sliding mode surface is that it can provide more flexibility when determining the sliding surface.

In the presence of the unmeasured states, adaptive sliding mode controller with a sliding mode observer was proposed. A nonlinear sliding mode observer, using only measured output information, was incorporated into the adaptive sliding mode control to estimate the unmeasured states, the angular velocity and gyroscope parameters. Moreover, adaptive sliding mode control for two axes' angular sensor was extended to a triaxial angular sensor. A novel concept for an adaptively controlled triaxial angular velocity sensor device that is able to detect rotation in three orthogonal axes and using a single vibrating mass was also proposed. In addition, a RBF neural network based adaptive sliding mode control for the state tracking control for triaxial angular velocity sensor was developed. The adaptive RBF neural network is incorporated into the adaptive sliding mode control scheme in the same Lyapunov framework to learn about the upper bound of model uncertainties and external disturbance. Meanwhile, all stiffness errors, damping terms and angular velocities can be consistently estimated in real time. Numerical simulations of MEMS gyroscope were investigated to show the effectiveness of the proposed advanced control schemes. It was shown that the proposed advanced controllers offer several advantages, such as consistent estimates of gyroscope parameters; including angular velocity and improved robustness to parameter variations and external disturbance.

Chapter 3 investigated the design of adaptive control for a MEMS gyroscope sensor. The dynamics model of the MEMS gyroscope sensor was developed and nondimensionized. Novel adaptive approach with a series-parallel on-line identifier was proposed and stability condition was established. Simulation results demonstrated the effectiveness of the proposed adaptive on-line identifier in identifying the gyroscope sensor parameters and angular rate. We should recognize that model uncertainties and external disturbances have not been considered in the proposed on-line adaptive identifier, as of yet. This should be compensated for in the real application, the next step is to incorporate the terms of model uncertainties and external disturbances into the online identifier, improving the robustness of the proposed method.

Chapter 4 developed the design of adaptive control for a MEMS gyroscope. The dynamics model of the MEMS gyroscope is developed and nondimensionized. A novel adaptive approach was proposed and a stability condition was established. Simulation results demonstrated the effectiveness of the proposed adaptive control techniques in identifying the gyroscope parameters and angular velocity.

Chapter 5 investigated the design of robust adaptive control for the gyroscope. A new robust adaptive controller was formulated for MEMS gyroscopes, having two unmatched oscillatory modes having had sufficient persistence of excitation to permit the identification of all gyroscope parameters, including the damping and stiffness coefficients and angular velocity. The proposed robust adaptive controller incorporates the capability to maintain a stable performance in the presence of model uncertainties and external disturbance. Numerical simulations show that the proposed robust adaptive control has satisfactory performance and robustness in the presence of model uncertainty and external disturbance.

Chapter 6 discussed the design of a model reference adaptive state feedback control with sliding mode property. A novel adaptive sliding mode controller with a proportional and integral sliding surface was proposed and developed. The stability of a closed-loop system could be established with the proposed adaptive sliding mode control structure and an integral sliding action. The controller proposed here used a novel sliding mode algorithm, consisting of a proportional and integral sliding surface. An adaptive sliding mode controller was derived to control the axes of the gyroscope and to estimate the unknown angular velocity. The proposed adaptive sliding mode control structure, with proportional and integral sliding action, could handle both matched and unmatched uncertainties and disturbance. The latter being true, provided that upper bounds for these uncertainties are available. Simulation results demonstrated that the use of the proposed proportional-integral sliding mode adaptive control technique was effective in estimating the gyroscope parameters and angular velocity in the presence of matched and unmatched disturbances.

Chapter 7 analyzed the design of a model reference adaptive state feedback sliding mode control for a MEMS angular velocity sensor. The controller proposed used a standard sliding mode algorithm. An adaptive sliding mode controller was derived to control the axes of the gyroscope and to estimate the unknown angular velocity. An adaptive law for the estimation of the unknown upper bound of the parameter uncertainties and external disturbance was derived. Furthermore, a smooth version of the adaptive sliding mode controller was used to reduce the control chattering. The proposed adaptive sliding mode control structure could establish the stability of a closed-loop system. The simulation results demonstrate that the use of the proposed sliding mode control technique was effective in estimating the gyroscope parameters and angular velocity in the presence of external disturbance and model uncertainties.

Chapter 8 presented an adaptive sliding mode controller with a proportional and integral sliding surface for a MEMS gyroscope. A nonlinear sliding mode observer, using only the position signal was incorporated into the adaptive sliding mode control algorithm. All unmeasured states, as well as the angular velocity and unknown gyroscope parameters, were consistently estimated. The combined observer-controller synthesis involves three steps: first a sliding mode controller was developed , assuming the availability of the state vector; second, a sliding mode observer of the state vector was designed to estimate the unmeasured states; and third, the sliding mode observer was combined with the proposed sliding mode controller, utilized the estimate instead of the true state vector. Simulations demonstrated the robustness of the proposed adaptive sliding mode controller with the sliding mode observer. It was shown that the angular velocity and gyroscope parameters could be consistently estimated in the presence of model uncertainty, external disturbance and unmeasured states.

Chapter 9 conducted the designs of an adaptive control and adaptive sliding mode control for a MEMS gyroscope. Novel adaptive approaches were proposed and Lyapunov stability conditions were established. The difference between these two adaptive approaches is that the external disturbance is considered in the derivation of adaptive sliding mode algorithm and the robustness can be improved. Simulation results demonstrated the effectiveness of the proposed adaptive control techniques in identifying the gyroscope parameters and angular velocity. Moreover, it could be concluded that adaptive sliding mode control had better robustness in the presence of external disturbance when compared with adaptive control.

Chapter 10 derived an adaptive sliding mode control for the state tracking control of a triaxial angular velocity sensor. A triaxial sensor substrate was adapted for use in measuring the acceleration and angular velocity of a moving body along three orthogonal axes. The controller drove the single mass along a controlled oscillation trajectory, removing the need for additional drive control. The proposed device included a single suspended mass that could move in three axles, having actuation and sensing elements in three orthogonal axes. The simulation was implemented to verify the effectiveness of the proposed adaptive sliding mode control for this triaxial angular velocity sensor.

Chapter 11 presented the RBF network based adaptive sliding mode control for the state tracking control for triaxial angular velocity sensor. The controller drives the single mass along a controlled oscillation trajectory, removing the need for additional drive control. An adaptive RBF neural network was used to learn about the upper bound of model uncertainties and external disturbance. The output of the neural network was used as a compensating parameter and the effects of the system uncertainties could be eliminated. The stability of the closed-loop system could be guaranteed with the proposed adaptive RBF sliding mode control strategy. The simulation was implemented to verify the effectiveness of the proposed adaptive RBF sliding mode control for this triaxial angular velocity sensor.

This book proposed an adaptive sliding mode controller with a proportional and integral sliding surface for a MEMS gyroscope in the presence of unmeasured states. In the near future, real time experiments of an adaptive sliding mode controller for the MEMS gyroscope should be performed, evaluating the effectiveness of the proposed adaptive sliding mode control algorithm. Detailed exploration of the convergence rate and transient behavior should be also discussed in the future. It should be noted that they depend on the adaptation gains and ratios between the two mismatched natural frequencies.

The value of sliding surface converges to zero as time goes on. The smooth sliding mode control's boundary layer is fixed in the simulation. It is necessary to adaptively adjust the smooth sliding mode control parameter or boundary layer's width, based on the changing of sliding surface. There are some relationships between boundary layers and sliding surfaces.

Some advanced control algorithms should be investigated to control the MEMS gyroscope.; for example, an adaptive sliding mode control with a fuzzy logic controller for the MEMS gyroscope or an adaptive backstepping control with sliding mode controller for the MEMS gyroscope. In the case of the time varying angular velocity of a MEMS gyroscope, polynomial approximation and other mathematical approximation methods can be combined with new adaptive control algorithm for a time varying system.

The real time implementation would require a high performance digital controller. A DSpace real time system or FPGA digital implementation can be investigated to control the MEMS Gyroscope. For the nano-position system and high speed implementation requirement, FPGA seems have great advantages over classic real time control algorithms because of its high speed ability. The high speed integrated circuit hardware descriptive language (VHDL) of the digital implementation will be designed to evaluate the adaptive control algorithm.

REFERENCES

[1] V. I. Utkin, "Variable structure systems with sliding modes," *IEEE trans. on Automatic Control*, vol. 22, pp. 212-222, 1988.

[2] V. I. Utkin, Sliding Modes in Control Optimization, Springer-Verlag, Berlin, 1992.

[3] Narendra, K. S., Annaswamy, A. M., 1989. Stable Adaptive Systems. Prentice-Hall, Englewood Cliffs, NJ.

[4] K. J Astrom, B. Wittenmark, Adaptive Control. Addison-Wesley Publishing, 1989.

[5] P. A. Ioannou and J. Sun, Robust Adaptive Control. Upper Saddle River, NJ: Prentice-Hall, 1998.

[6] G. Tao, Adaptive Control Design and Analysis, John Wiley and Sons Inc., London, England, 2003.

[7] T. H. Lee, K. K. Tan, M. W. Lee, "A variable structure-augmented adaptive controller for a gyro mirror line of light stabilization platform," *Mechatronics*, vol. 8, pp.48-84, 1998.

[8] Y. P. Wang, A. Sinha, "Adaptive sliding mode control algorithm for a microgravity isolation system,"*Acta Astronautica*, vol. 43(8-8), pp. 388-384, 1998.

[9] G. Song, R. Mukherjee, "A comparative study of conventional nonsmooth time-invariant and smooth time-varying robust compensator," *IEEE Trans. on Control System Technology*, vol. 8(4), pp. 581-588, 1998.

[10] Y. Sam, J. H. Osman and M. R. Ghani, "A class of proportional-integral sliding mode control with application to active suspension system," *Systems and Control Letters*, vol. 51, pp. 218–224. 2004.

[11] C. Chou, C. Cheng , "A decentralized model reference adaptive variable structure controller for large-scale time-varying delay systems," *IEEE Trans. on Automatic Control,* vol. 48(8), pp. 1213-1218, 2003.

[12] F. Lin, S. Chiu and K. Shu, " Novel sliding mode controller for synchronous motor drive," *IEEE Trans. on Aerospace and Electronic Systems*, vol. 34(2), pp. 532-541, 1998.

[13] L. Hsu, A. D. Araujo, R. Costa, "Analysis and design of I/O based variable structure adaptive control," *IEEE Trans.on Automatic control*, vol. 39(1), pp. 4 -21, 1994.

[14] S. Park, and R. Horowitz, "Adaptive Control for the Conventional Mode of Operation of MEMS Gyroscopes," Journal of *Microelectromechanical Systems*, vol. 12(1), pp. 101-108, 2003.

[15] A. M. Shkel, R. Horowitz, A. A. Seshia, S. Park, and R. T. Howa, "Dynamics and control of micromachined gyroscopes, " *Proceedings of the 1999 American Control Conference*, vol. 3, pp. 2119-2124, 1999.

[16] R. Leland, "Adaptive mode tuning for vibrational gyroscopes", IEEE *Trans. on Control System Technology*, vol .11(2), pp. 242-248, 2003.

[17] R. Leland; Y. Lipkin, A. Highsmith, "Adaptive oscillator control for a vibrational gyroscope," *Proceedings of 2003 American Control Conference*, vol. 4, pp.3348-3352, 2003.

[18] R. Leland, "Lyapunov based adaptive control of a MEMS gyroscope," *Proceedings of 2002 American Control Conference*, vol. 5, pp.3885-3880, 2002.

[19] L.Dong, R. Leland, "The adaptive control system of a MEMS gyroscope with time-varying rotation rate," Proceedings of the 2005 *American Control Conference*, vol. 5, pp.3592-3598, 2005.

[20] C. Batur, T. Sreeramreddy and Q. Khasawneh, "Sliding mode control of a simulated MEMS gyroscope," *Proceedings of the 2005 American Control Conference*, vol. 2, pp. 4180 – 4185, 2005.

[21] Hameed. S, Jagannathan, "Adaptive force-balancing control of MEMS gyroscope with actuator limits," *Proceedings of the 2004 American Control Conference*, vol. 2, pp.1882-1888, 2004.

[22] S. Park, Adaptive control strategies for MEMS gyroscope, Ph.D. book, University of California, Berkeley, 2000.

[23] S. Park, R. Horowitz, "New adaptive mode of operation for MEMS Gyroscopes," ASME Transaction of Dynamic Systems, *Measurement and Control*, vol. 128, pp.800-810, 2004.

[24] S. Park, R. Horowitz, "Discrete time adaptive control for a MEMS gyroscope, " *International Journal of Adaptive Control and Signal Processing* , vol. 19(8), pp.485-503, 2005.

[25] F. Esfandiari, H. K. Khalil, "Output feedback stabilization of fully linearizable systems," *Int. J. Control*, 58, pp.1008-1038, 1992.

[26] M Krstic, I. Kanellallakopoulos and P. Kokotovic, Nonlinear and Adaptive Control Design, John Wiley and Sons Inc., 1995.

[27] J.-J. E. Slotine, J. K. Hedrick and E. A. Misawa, "On sliding observers for nonlinear systems," *ASME Transaction Dynamic Systems, Measurement and Control*, vol. 109(3), pp. 245–252, 1988.

[28] B. L. Walcott and S. H. Zak, "Combined observer controller synthesis for Uncertain Dynamical Systems with Applications," *IEEE Trans. Syst. Man. Cyberb.*, vol. 18(2), pp. 88-104, 1988.

[29] K. D. Young, Variable Structure Control for Robotics and Aerospace Applications, Elsevier Science Publishers, 1992.

[30] C. Batur and L. Zhang, "Sliding mode observer and controller design for a hydraulic motion control system," *Proceedings of the 2003 American Control Conference*, pp. 1821 – 1828, 2003.

[31] C. Edwards and S. K. Spurgeon, "Sliding mode output tracking with application to a multivariable high temperature furnace problem,"*International Journal of Robust and Nonlinear Control*, vol. 8, pp. 338–351, 1998.

[32] C. Edwards and S. K. Spurgeon, "Robust output tracking using a sliding mode controller/observer scheme," *International Journal of Control*, vol. 84, pp. 988–983, 1998.

[33] C. Edwards and S. K. Spurgeon, Sliding Mode Control: Theory and Applications. Basingstoke, U. K.: Taylor and Francis, 1998.

[34] J. Han. "A class of extended state observers for uncertain systems," Journal of Control and Decision, vol.10(1), pp. 85-88, 1995.

[35] J. T. Moura, H. Elmali, and N. Olgac, "Sliding Mode Control with Sliding Perturbation Observer," *ASME Transaction Dynamic Systems, Measurement and Control*, vol. 119(4), pp. 858–885, 1998.

[36] G. K. Fedder, Simulation of Microelectromechanical System, Ph.D. book, University of California, Berkeley, 1994.

[37] G. Wheeler, C.Y. Su, and Y. Stepanenko, "A sliding mode controller with improved adaptation laws for the upper bounds on the norm of uncertainties," *Automatica*, vol. 34(12), pp. 1858-1881, 1998.

[38] Y. Stepanenko, Y. Cao, and C. Y. Su, "Variable structure control of robotic manipulators with PID sliding surfaces," *International Journal of Robust and Nonlinear Control*, vol. 8(1), pp. 89-90, 1998.

[39] C.Y. Su, Y. Stepanenko, and T.P. Leung, "Combined adaptive and variable structure control for constrained robots," *Automatica*, vol. 31(3), pp. 483-488, 1995.

[40] C.Y. Su, and T.P. Leung, "A sliding mode controller with bound-estimation for robot manipulator," *IEEE Transactions on Robotics and Automation*, vol. 9(2), pp. 208-214, 1993.

[41] C.Y. Su, T.P Leung, and Y. Stepanenko, "Real-time implementation of regressor based sliding mode control scheme for robot manipulators," *IEEE Transactions on Industrial Electronics*, vol. 40(1), pp. 81-89, 1993.

[42] Q. Hu, G. Ma, "Control of three-axis stabilized flexible spacecraft using variable structure strategies subject to input nonlinearities," Journal of *Vibration and Control*, vol. 12(8), pp. 859-881, 2008.

[43] D. S. Yoo and M. J. Chung, "A variable structure control with simple adaptation laws for upper bounds on the norm of the uncertainties," *IEEE Transactions on Automatic Control*, vol. 38(3), pp. 159-185, 1992.

[44] G. Song, L. Cai, Y. Wang and R. W. Longman, "A sliding mode based smooth adaptive robust controller for friction compensation," *Int. J. Robust and Nonlinear Control*,vol. 8, pp.825-839, 1998.

[45] J. Yan, K. Shyu and J. Lin, "Adaptive variable structure control for uncertain chaotic systems containing dead-zone nonlinearity," *Journal of Intelligent and Robotic Systems,* vol. 25, pp. 348-355, 2005.

[46] F. Pourboghrat, G. Vlastos, "Model reference adaptive sliding control for linear systems," *Computers and Electrical Engineering*, vol. 29, pp. 381-384, 2002.

[47] Z. Man, D. Habibi, "A robust adaptive sliding-mode control for rigid robotic manipulators with arbitrary bounded input disturbance," *Journal of Intelligent and Robotic Systems*, vol.18, pp.381-388, 1998.

[48] C. Chou, C. Cheng, "Design of adaptive variable structure controls for perturbed time-varying state delay systems," *Journal of the Franklin Institute*, vol. 338, pp.35-48, 2001.

[49] Y. Roh, J. Oh, "Sliding mode control with uncertainty adaptation for uncertain input-delay systems," *Int. J. Control*, vol. 83(13), pp.1255-1280, 2000.

[50] M. Abe, E. Shinohara, K. Hasegawa, S. Murata, and M. Esashi, "Trident-type tuning fork silicon gyroscope by the phase difference

detection," The 13th Annual International Conference on Micro Electro Mechanical Systems, pp. 508-513, 2000.

[51] C. Acar, S. Eler, and A. Shkel, "Concept, implementation, and control of wide bandwidth mems gyroscopes," *Proceedings of the 2001 American Control Conference*, vol. 2, pp. 1229-1234, 2001.

[52] C. Acar and A. Shkel, "Distributed-mass micromachined gyroscopes: demonstration of drive-mode bandwidth enhancement," *Proceedings of the 2004 Electronic Components and Technology*. ECTC '04, vol.1, pp. 884-882, 2004.

[53] S. E. Alper and T. Akin, "A symmetric surface micromachined gyroscope with decoupled oscillation modes," *Sensors and Actuators A: Physical*, vol. 98, pp. 348-358, 2002.

[54] T. Gabrielson, "Mechanical-thermal noise in micromachined acoustic and vibration sensors," *IEEE Transactions on Electron Devices*," vol. 40(5), pp. 903-909, 1993.

[55] B. Gallacher, J. Hedley, J. Burdess, A. Harris, A. Rickard, and D. King, "Electrostatic correction of structural imperfections present in a microring gyroscope," *Journal of Microelectromechanical Systems*, vol. 14(2), pp. 221-234, 2005.

[56] X. Jiang, J. Seeger, M. Kraft, and B. Boser, "A monolithic surface micromachined z-axis gyroscope with digital output," *Proceedings of the 2000 Symposium on Application In VLSI Circuits*, pp. 18-19, 2000.

[57] J. John, C. Jakob, T. Vinay, "Phase differential angular velocity sensor-concept and analysis," *IEEE Sensors Journal*, vol. 4(4), pp. 481-488, 2004.

[58] J. John and T. Vinay, "Novel concept of a single mass adaptively controlled triaxial angular velocity sensor," *Sensors Journal IEEE*, vol. 8(3), pp.588-595, 2008.

[59] D. Keymeulen, W. Fink, M. Ferguson, C. Peay, B. Oks, R. Terrile, and K. Yee, "Tuning of mems devices using evolutionary computation and open-loop frequency response," 2005 *IEEE Conference In Aerospace* , pp. 1-8, 2005.

[60] S. Kim, B. Lee, J. Lee, and K. Chun, "A gyroscope array with linked-beam structure," The 14th IEEE International Conference on Micro Electro Mechanical Systems, pp.30-33, 2001.

[61] R. L. Kubena, D. J. Vickers-Kirby, R. J. Joyce, F. P. Stratton, and D. T. Chang, "A new tunneling-based sensor for inertial rotation rate measurements," *Sensors and Actuators A: Physical*, vol. 83(1-3), pp.109-118, 2000.

[62] Y. Liang, T. Zhao, Y. Xu, and S. Boh, "A cmos fully-integrated low-voltage vibratory microgyroscope, " *Proceedings of IEEE Region 10 International Conference on Electrical and Electronic Technology*, vol. 2, pp.825-828, 2001.

[63] R. Closkey, S. Gibson, and J. Hui, "System identification of a mems gyroscope," *ASME. Transactions of Dynamic Systems, Measurement and Control*, vol. 123(2), pp 201-210, 2001.

[64] Y. Mochida, M. Tamura, and K. Ohwada, "A micromachined vibrating rate gyroscope with indepen-dent beams for the drive and detection modes," *IEEE International Conference on Micro Electro Mechanical Systems*, pp. 818-823, 1999.

[65] C. Painter and A. Shkel, "Structural and thermal analysis of a mems angular gyroscope," *Proceedings of the SPIE - The International Society for Optical Engineering*, vol. 4334, pp. 88-94, 2001.

[66] K. Park, C. Lee, Y. Oh, and Y. Cho, "Laterally oscillated and force-balanced micro vibratory rate gyroscope supported by hook shape springs," *Proceedings of IEEE 10th International Workshop on Micro Electro Mechanical Systems*, pp.494-499. 1998.

[67] P. Qi Lin, Stern, "Analysis of a correlation fillter for thermal noise reduction in a mems gyroscope," *Proceedings of the 34th Southeastern Symposium on System Theory*, pp.198-203, 2002.

[68] A. Seshia, R. Howe, and S. Montague, "An integrated micro-electromechanical resonant output gyroscope," *IEEE International Conference on Micro Electro Mechanical Systems*, pp. 822-828, 2002.

[69] T. Tang, R. Gutierrez, "A packaged silicon mems vibratory gyroscope for microspacecraft," *Proceedings of the IEEE 10th International Workshop on Micro Electro Mechanical Systems*, pp. 500-505, 1998.

[70] C.-h.Wang and X. Huang, "Application of wavelet packet analysis in the de-noising of mems vibrating gyro," *Proceedings of the 2004 Position Location and Navigation Symposium*, pp.129-132, 2004.

[71] G. Wu, Z. Xiao, Z. Li, and Y. Hao, "Design and fabrication for inertial micro sensors," *Proceedings of 5th International Conference on Solid-State and Integrated Circuit Technology*, pp. 903-909, 1998.

[72] H. Yang, M. Bao, H. Yin, and S. Shen, "A novel bulk micromachined gyroscope based on a rectangular beam-mass structure," *Sensors and Actuators A: Physical*, vol. 98(2-3), pp.145-151, 2002.

[73] N. Yazdi, F. Ayazi, and K. Naja, "Micromachined inertial sensors," *Proceedings of the IEEE*, vol. 88(8), pp. 1840-1859, 1998.

[74] M. H. Salah, M. McIntyre, D. M. Dawson, and J. R. Wagner, "Time-varying angular velocity Sensing for a MEMS Z-Axis Gyroscope," *Proceedings of the IEEE Conference on Decision and Control*, pp. 581-588, 2008.

[75] T. Juneau, A.P. Pisano, J. H. Smith, "Dual axis operation of a micromachined rate gyroscope," 1998 *International Conference on Solid State Sensors and Actuators*, vol. 2, pp.883 – 888, 1998.

[76] R. Oboe, R. Antonello, E. Lasalandra, G.Durante, and L. Prandi, "Control of a *Z*-Axis MEMS Vibrational Gyroscope, " *IEEE/ASME Transaction on Mechatronics*, vol. 10(4), pp.384-380, 2005.

[77] C. Liu, Foundations of MEMS, Upper Saddle River, NJ, Pearson Prentice Hall, 2008

[78] S. D. Senturia, Microsystem Design, Boston: Kluwer Academic Publishers, 2001.

[79] B.R. Andrievsky, A.A. Stotsky, A.L. Fradkov, "Velocity gradient algorithms in control and adaptation problems: A survey," *Automation Remote Control*, vol. 12, pp.1533-1584, 1988.

[80] B.R. Andrievsky, A.L. Fradkov, A.A. Stotsky, "Shunt compensation for indirect sliding-mode adaptive control," 13th IFAC World Congress, San Francisco, pp. 193-198, 1998.

[81] A.L. Fradkov, B.R. Andrievsky, "Combined adaptive controller for UAV guidance," *European J. of Control*, vol. 11 (1), pp.81-89, 2005.

[82] A. L. Fradkov, "Speed-gradient Scheme and its Applications in Adaptive Control," Automation. Remote Control, vol. 40(9), pp.1333-1342, 1989.

[83] B. Acar, "Robust micromachined vibratory gyroscopes," Ph.D. book, University of California, Irvine, 2004.

[84] Q. Zheng, L. Dong, Z. Gao, "Control and rotation rate estimation of vibrational MEMS gyroscopes," pp. 118-123, 18th *IEEE International Conference on Control Applications*, 2008.

[85] Z. Feng, M. Fan, "Adaptive input estimation methods for improving the bandwidth of microgyroscopes," vol.8(4), pp.582-588, *IEEE Sensors Journal*, 2008.

[86] J. Fei, C. Batur, "Adaptive sliding mode control with sliding mode observer design for a MEMS vibratory gyroscope," *Proc. of IMechE, Part I, Journal of System and Control Engineering*, vol. 222(8), pp. 839-849, 2008.

[87] J. Fei, C. Batur, "Robust adaptive control for a MEMS vibratory gyroscope," *International Journal of Advanced Manufacturing Technology*, vol. 42(3), pp. 293-300, 2009.

[88] J. Fei, C. Batur, "A novel adaptive sliding mode control with application to MEMS gyroscope," *ISA Transactions*, vol. 48(1), pp.83-88, 2009.

[89] J. Fei, "Robust adaptive vibration tracking control for a MEMS vibratory gyroscope with bound estimation," *IET Control Theory and Application*, vol. 4(8), pp. 1019-1028. 2010.

[90] Y. Guo, P. Woo, "An adaptive fuzzy sliding mode controller for robotic manipulators," *IEEE Transactions on Systems, Man and Cybernetics-Part* A, vol. 33(2), pp. 149-159. 2004.

[91] R. J. Wai, "Fuzzy sliding-mode control using adaptive tuning technique," *IEEE Transactions on Industrial Electronics*, vol. 54(1), pp. 588–594, 2008.

[92] S. Lin, Y Chen, "RBF network based sliding mode control," *IEEE Transactions on Systems, Man and Cybernetics,* pp.1958-1981, 1994.

[93] N. Sadati, R. Ghadami, "Adaptive multi-model sliding mode control of robotic manipulators using soft computing," *Neurocomputing,* vol. 81(2), pp. 2802–2810, 2008.

[94] B. S. Park, S.J. Yoo, J. B. Park, Y. H. Choi, "Adaptive neural sliding mode control of nonholonomic wheeled mobile robots with model uncertainty," *IEEE Transactions on Control System Technology, vol.* 18(1), pp. 208 – 214,2009.

[95] M. Lee, Y. Choi, "An adaptive neucontroller using RBFN for robot manipulators," *IEEE Transactions on Industrial Electronics*, vol. 51(3), pp. 811–818, 2004.

[96] X.Yu , O. Kaynak, "Sliding-mode control with soft computing: A Survey," *IEEE Transactions on Industrial Electronics*, vol. 58(9), pp. 3285 – 3285, 2009.

[97] J. Liu, MATLAB Simulation for Sliding Mode Control. Tsinghua University Press, Beijing, 2005.

[98] J. Fei, F. Chowdhury, "Robust adaptive controller for triaxial angular velocity sensor," *International Journal of Innovative Computing, Information and Control*, vol. 7(6), pp.2439-2448, 2010.

[99] J. Fei, C. Batur, "A class of adaptive sliding mode controller with integral sliding surface, Proc. of IMechE, Part I," *Journal of System and Control Engineering*, 223(17), pp. 989-999.

INDEX

A

B

C

D

T

U

V

W

Y